Lecture Notes in Energy

Volume 72

Lecture Notes in Energy (LNE) is a series that reports on new developments in the study of energy: from science and engineering to the analysis of energy policy. The series' scope includes but is not limited to, renewable and green energy, nuclear, fossil fuels and carbon capture, energy systems, energy storage and harvesting, batteries and fuel cells, power systems, energy efficiency, energy in buildings, energy policy, as well as energy-related topics in economics, management and transportation. Books published in LNE are original and timely and bridge between advanced textbooks and the forefront of research. Readers of LNE include postgraduate students and non-specialist researchers wishing to gain an accessible introduction to a field of research as well as professionals and researchers with a need for an up-to-date reference book on a well-defined topic. The series publishes single- and multi-authored volumes as well as advanced textbooks. **Indexed in Scopus and EI Compendex** The Springer Energy board welcomes your book proposal. Please get in touch with the series via Anthony Doyle, Executive Editor, Springer (anthony.doyle@springer.com).

More information about this series at http://www.springer.com/series/8874

Miguel Moreira da Silva

Power and Gas Asset Management

Regulation, Planning and Operation of Digital Energy Systems

Springer

Miguel Moreira da Silva
Porto, Portugal

ISSN 2195-1284 ISSN 2195-1292 (electronic)
Lecture Notes in Energy
ISBN 978-3-030-36199-0 ISBN 978-3-030-36200-3 (eBook)
https://doi.org/10.1007/978-3-030-36200-3

This Springer imprint is published by the registered company Springer Nature Switzerland AG
The registered company address is: Gewerbestrasse 11, 6330 Cham, Switzerland

To my Mother

Foreword

I initially met Miguel Moreira da Silva when he was a PhD student in the MIT Portugal program investigating energy storage and electric mobility for sustainable energy systems. A decade later, I'm thrilled to introduce Dr. Miguel Moreira da Silva's excellent book on power and gas asset management. Miguel delivers a political, regulatory, and technology framework to support energy asset management transformation rooted in digital systems, data-centric organizations, and advanced analytics. With energy decarbonization, liberalization, and digitalization as its drivers, this book brings in asset management as a critical methodology for both the scientific community and industry stakeholders.

This book will be crucial for educating future energy system asset managers, covering the dynamics of the electricity and gas sectors. It is important that asset managers understand how energy utilities are impacted by international and national policies, as well as by technology disruption. Additionally, Dr. Moreira da Silva discusses the impact of regulation on asset replacement and maintenance policies. Incentive and performance-based approaches are gaining traction to accelerate risk-based asset management and predictive maintenance.

Risk and condition-based maintenance require a comprehensive asset management transformation. Besides technology investment, Miguel argues the utility must undergo a bold cultural and organizational transformation. This book proposes a target operating model—"Portfolio Management"—to capture the full potential of risk-based asset management. The model advocated by Miguel discusses a set of three transformative actions, namely: i) internal reorganization of teams and processes; ii) design of coherent decision-making models (aligned with the regulatory regime); and iii) asset digitalization and advanced data analytics.

When it comes to decision models and advanced analytics, this book provides innovative approaches—probabilistic and deterministic—for risk modeling. Miguel formulates a risk model based on asset fragility and criticality, coupled with practical examples. Learning algorithms are also proposed as a promising approach to handle predictive maintenance problems.

This text will be an important educational tool for future generations of asset managers.

Dava J. Newman
Former NASA Deputy Administrator (2015–2017)
Apollo Program Professor of Astronautics at MIT
Cambridge, MA, USA

Contents

About the Author

Miguel Moreira da Silva (Portugal, 1983) is a senior engineer and professor in energy systems. He holds a Doctor degree in Sustainable Energy Systems from the MIT Portugal program at the Faculty of Engineering, University of Porto (FEUP) and is a chartered Electrical and Computer Engineer (FEUP). In 2013, Moreira da Silva was selected one of the 100 "Future Energy Leaders", by the World Energy Council, and in 2019 has joined the Global Panel of the MIT Technology Review.

Moreira da Silva has been working in the energy industry for nearly 15 years, close to board of directors and government officials, and covering all the value chain. For seven years he performed engineering and management roles at REN[1], where he led the power and gas data and analytics, electricity asset management, and R&D departments. Previously he served as Deputy to the Minister for Environment, at the Government of Portugal, holding the cabinet's responsibility for drafting and negotiating national policies for sustainable development, energy decarbonization and water sector regulation. Moreira da Silva has also worked in energy technology corporations, namely Itron and Iskraemeco. He is now director for strategic planning and people analytics at the retail subsidiary of Sonae group, and is still member of the jury of REN Scientific Award.

Besides the business experience, for five years Moreira da Silva lectured Power Systems Engineering, at IST—University of Lisbon. He has also been collaborating with INESC-TEC in advance analytics training for energy industry professionals. Within the scope of his research activity, Moreira da Silva has published and presented several works in the fields of energy planning, asset management, energy storage, electric mobility and smart grids.

[1]Portugal's power and gas transmission system operator, and concessionaire of one gas distribution grid.

List of Figures

List of Tables

Chapter 1
Introduction

1.1 Drivers for Asset Management

Global energy systems are undergoing an unprecedented transformation era. Renewable energy sources (RES) are soaring from West to East. Consumers grip energy efficiency levers, showcasing an utter environment awareness of energy use impacts. Energy infrastructures are starting a digitalization journey, road to customer-centric services, dynamic markets, efficient operation and maintenance (O&M), and people and goods safety. After a first wave of energy sector liberalization and privatization, there is now an upsurge of energy portfolio shift, merger and acquisitions, and innovative regulatory regimes, tackling not only investment effectiveness and operation efficiency, but also quality of service, research & development and environment protection.

This portrait will be drawn, globally, at different paces and with diverse solutions. In Europe and some regions of North and South America, following an early momentum of investment in renewable energy sources and grid expansion, in the power sector, and development of national-wide pipeline infrastructure, in the natural gas side, energy utilities are facing new technical and regulatory challenges. The upcoming energy ecosystem will be characterized by a steady RES investment trend (as a contrast to the previous exponential development); a plateau of the annual load growth (due to the rollout of energy efficient technologies, consumers' behavioral change and self-consumption systems); and mature transmission and distribution grids (having the transnational interconnections as the remaining bulk investment). In this context, one expects an adaptation of the regulatory models to the new energy landscape.

Capital-recovery remuneration and cost of service approaches are deemed adequate in periods of fast development of power and gas infrastructures (to meet adequacy and service quality goals). On the other hand, for consolidated and mature grids, the regulatory periods may be adjusted (longer than for capital-intensive regimes) together with different remuneration rules. Instead of addressing capital

M. Moreira da Silva, *Power and Gas Asset Management*, Lecture Notes in Energy 72, https://doi.org/10.1007/978-3-030-36200-3_1

and operational costs separately, future regulation will likely gather both investment and maintenance activities in the same financial envelope, under a life-cycle philosophy. Regulatory models in such situation could be based on total expenditures (TOTEX) targets, with reliability and security of supply key performance indicators (KPIs), and innovation objectives.

Developing economies, on their side, are rocketing investment in low carbon technologies, coupled with fossil fuel power plants decommissioning. This technology swap implies the generation and transmission asset end-of-life management, coordinated with the forthcoming generation capacity, grid growth and reinforcement. Still digital energy systems will thrive worldwide, regardless the energy decarbonization stage.

Regulation will be surely influenced by technology breakthroughs, as a mean to optimize processes and deliver innovative services. Besides the technical findings related with RES, battery energy storage, grid automation (including static relays and advanced metering) and power electronics, electricity technologies are still rooted on the same fundamentals: Faraday's law of induction; Kirchoff's current and voltage laws; and Ohm's laws. Like in the late XIX century, the electricity value chain encloses generators, transmission lines, power transformers and breakers. A parallel analysis can be made to natural gas. As in its deployment, in early XX century, natural gas systems are still encompassed by pipelines, boilers and diaphragm meters, and ruled by the conservation of mass, Reynolds transport theorem and Navier-Stokes equations.

In the next few years both the electricity and natural gas systems will observe a holistic transformation of their processes and assets, as a result of the rollout of digital technologies and learning algorithms. From electricity generation to distribution, and from natural gas upstream to downstream operations, digitalization will reshape century business models. Electronic, real-time, two-way, cloud-based and open source technologies will replace analogic, offline, one-way, in-house and proprietary solutions.

The aforementioned context sculpts the future energy industry landscape, where asset management will play an underlying role. If designed and executed in a proper way, asset management delivers a set of benefits, namely (Hastings 2014; The Institute of Asset Management 2008):

- enhanced customer satisfaction from improved performance and control of product/service delivery;
- improved health, safety and environmental performance;
- optimized return on investment and/or growth;
- systematic approach to asset-based decisions;
- adequate logistic support throughout the asset life cycle;
- effective internal processes, capable of demonstrating legal, regulatory and statutory compliance;
- improved risk management, corporate governance and reputation;
- workforce training and talent development.

Yet this book does not seek to address, conceptually, the asset management systems, processes and activities. This publication aims at delivering the political, regulatory and technological framework to support the energy asset management transformation, rooted on digital systems, data-centric organization and advanced analytics. It is hence a research on end-to-end asset management towards sustainable, efficient and innovative power and gas systems.

1.2 Book Outline

This book provides an overview on the energy industry, regulation and business models, before approaching the asset management topic. In truth, the energy asset management is only effective if properly framed with political, technological and regulatory contexts. In view of that, the first chapter dives into the details of climate and energy policies. The asset manager role is thoroughly impacted by international and national energy agenda, from RES investment to security of supply. Despite sharing a set of energy transition drivers, power and gas systems embody sectoral idiosyncrasies. In view of that, a state of art is provided to each energy system, by addressing sustainable and flexible systems, for the power sector, and geopolitics, interconnections and LNG, for gas business.

Being energy systems dependent of regulatory regimes and market models, chapter two brings the fundamentals of regulation theory. Furthermore, power and gas value chains are characterized, and a recipe is drawn for sectoral liberalization and competition. To complement the analysis, this chapter introduces a detailed description of power generation security, economics and market operations. Regulation of energy transmission and distribution is also investigated, including a review on regulatory models coupled by an industry benchmark. LNG is similarly subject to an outline on the concerned operations and business services.

Having made an overview on energy policies, trends, technology and regulation, then in chapter four asset management is explored. Firstly, one proposes the mission, values, strategy and objectives of energy asset management. A state of art is subsequently developed for the possible operating models. After presenting the various types of maintenance policies, eventually a target operating model is described. In this context, adequate governance and organization are advocated towards risk-based investment and maintenance. Finally, chapter four presents a roadmap for digital transformation (including a business case) and a strategy for building a data-centric utility.

At last, chapter five introduces the fundamentals of decision aid methodologies and risk modelling. Asset fragility and criticality indices are formulated and complemented with practical asset-related use cases. Risk indices are based on different approaches, including deterministic and probabilistic models, as well as learning algorithms.

References

Hastings NAJ (2014) Physical asset management with an introduction to ISO55000, 2nd edn. Springer, Berlin

The Institute of Asset Management (2008) PAS55, Part 1: specification for the optimized management of physical assets

Chapter 2
Energy Industry Landscape

2.1 Climate and Energy Agenda

2.1.1 The Scientific Evidence of Climate Change

Changes in the atmospheric abundance of greenhouse gases (GHG) and aerosols, in solar radiation and in land surface properties modify the energy balance of the climate system. These changes are expressed in terms of radiative forcing, which is used to compare how a range of human and natural factors drive warming or cooling influences on global climate (IPCC 2007).

The stocks of GHG in the atmosphere are rising, as a result of human activity, being carbon dioxide (CO_2) the most important anthropogenic GHG (Stern 2006). Indeed, CO_2 concentration in the atmosphere has been considerably increasing in the last century, compared to levels in the pre-industrial era, i.e. 280 parts per million (ppm). In 2016, the average concentration of CO_2 was roughly 403 ppm, which is circa 40% higher than in the mid-1800s. In addition, methane and nitrous oxide emissions have also risen (IEA 2017a). As the Intergovernmental Panel on Climate Change (IPPC) stated in (IPCC 2007), the main cause for carbon dioxide increase since the pre-industrial period stands for fossil fuel use and land-use change (with a smaller impact). Stern (2006) also highlights the increasing risks of serious, irreversible impacts from climate change, linked with business-as-usual paths for emissions. It is believed that the level of greenhouse gases in the atmosphere would double pre-industrial levels by 2050 (i.e. 550 ppm) and would keep rising thereafter, for a scenario without increases on the annual flow of emissions beyond today's rate.

According to IPCC (2018), human activities may have caused about 1.0 °C of global warming above pre-industrial levels. In fact, global warming is likely to reach 1.5 °C between 2030 and 2052 if it continues to increase at the current rate (IPCC 2018).

M. Moreira da Silva, *Power and Gas Asset Management*, Lecture Notes in Energy 72, https://doi.org/10.1007/978-3-030-36200-3_2

In Fig. 2.1 one can observe the evolution of both the monthly global mean surface temperature (grey line) and the estimated anthropogenic warming to date (solid orange line) and likely range (orange shading). For 2017 onwards, IPCC provides the central estimate (orange dashed arrow) and likely range of the time (orange error bar) at which 1.5 °C is reached if the current rate of warming continues. The grey plume shows the likely range of warming responses in which net CO_2 emissions decay in a straight line from 2020 to achieve net zero in 2055 and net non-CO_2 radiative increases to 2030 and then reduces. The blue plume depicts the response to faster CO_2 emissions abatement, reaching net zero in 2040, whereas the purple plume provides the response to net CO_2 emissions declining to zero in 2055, with net non-CO_2 forcing remaining constant after 2030 (IPCC 2018) (Figs. 2.2 and 2.3).

According to Stern (2006), stabilisation of GHG in the atmosphere is feasible and compatible with economic growth. The developed countries have noticed a decrease on the responsiveness of emissions to income growth, owing to changes in energy technologies and the structure of economies (Stern 2006). The fact is that it is possible to decarbonise both developed and developing economies, while tackling economic growth in both.

Carbon emissions stabilisation requires a slump of annual emissions to the level that balances the Earth's natural capacity to remove GHG from the atmosphere. Stern (2006) foresees that the annual costs for a stabilisation at 500–550 ppm of CO_2 concentration would be around 1% of GDP by 2050. The author reckons the GHG emissions can be cut in four ways: **a** reducing demand for emissions-intensive goods and services; **b** increased efficiency, which can save both money and emissions; **c** action on non-energy emissions, such as avoiding deforestation; and **d** switching to lower-carbon technologies for power, heat and transport.

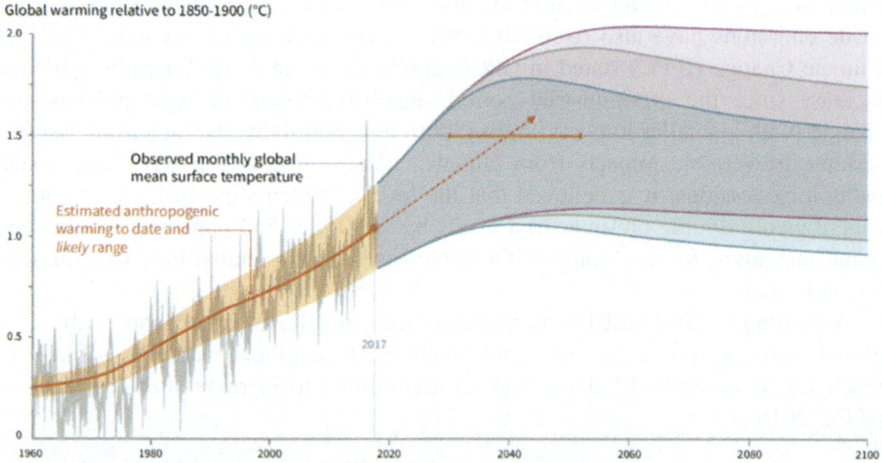

Fig. 2.1 Historical global warming and future pathways (IPCC 2018)

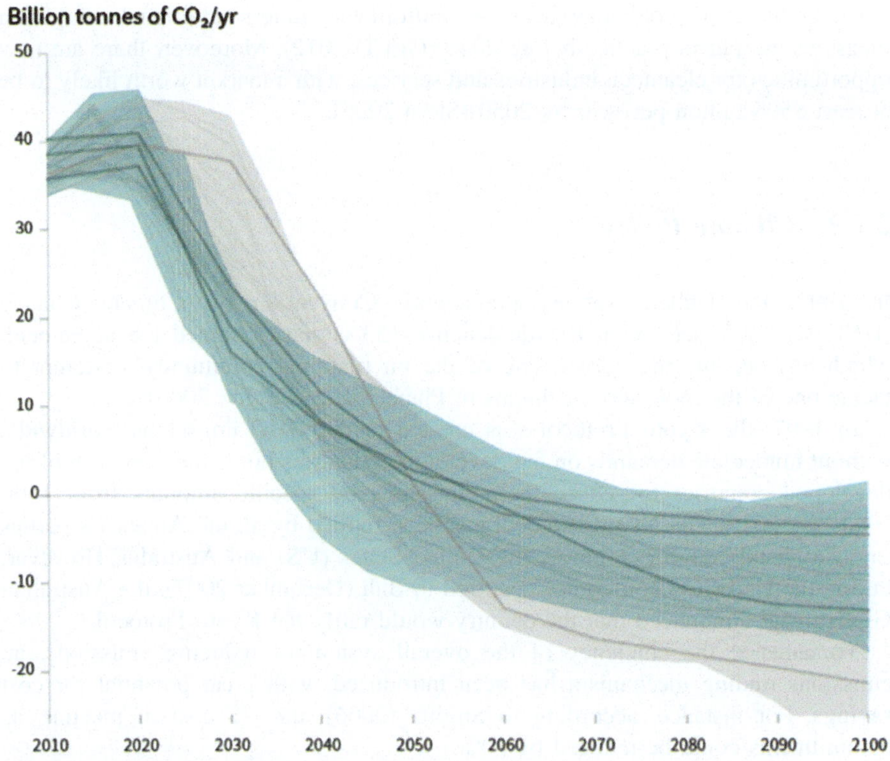

Fig. 2.2 Comparison of global CO_2 emissions from different strategies (IPCC 2018)

Fig. 2.3 Cumulative CO_2 emissions in pathways reaching net zero in 2055 (grey line) and 2040 (blue line) (IPCC 2018)

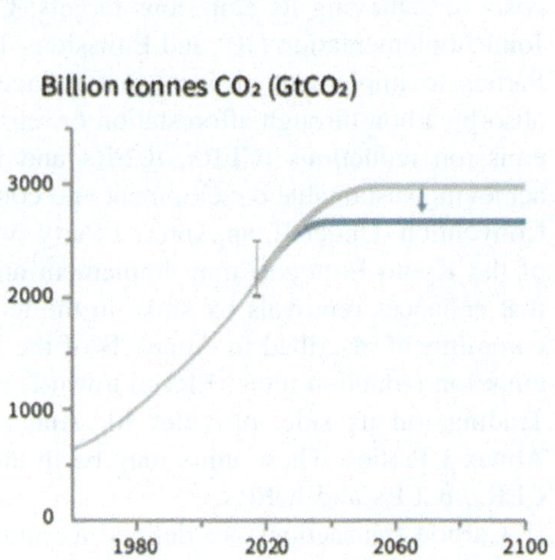

A key driver for low carbon economy consists in the high price of delay, which would lead to acceptance of more climate change and higher mitigation costs. As OECD (2012) highlighted, the costs of delayed action may be so high that reaching the 2 °C goal may become unaffordable, unless additional mitigation measures are put in practice before 2020 (OECD 2012). Moreover, there are new opportunities for cleantech industries and services, with a market worth likely to be at least $500 billion per year by 2050 (Stern 2006).

2.1.2 Climate Policy

In 1992, the United Nations Framework Convention on Climate Change (UNFCCC) was set up, in Rio de Janeiro. 153 countries signed the agreement, which has become the cornerstone of the international community's attempt to tackle one of the most serious threats to Planet Earth (Stiglitz 2006).

In 1997, the Kyoto Protocol was created to cut GHG emissions worldwide, without immediate demands on the developing countries but calling on each of the developed countries to reduce their emissions by specific amounts from 1990 levels, by 2012. The Kyoto Protocol has been ratified by all the Annex I's parties (industrialized countries), unless the United States (US) and Australia. However, during the UNFCCC conference occurred in Bali (December 2007), the Australian Government announced that the country would ratify the Kyoto Protocol.

To enhance the efficiency of the overall system of reducing emissions, an emissions trading mechanism has been introduced, with great potential for cost savings. For instance, according to Stiglitz (2006), the US cost of meeting its commitments could be reduced by 60%.

The Kyoto Protocol defined three "flexibility mechanisms" to lower the overall costs of achieving its emissions targets: Clean Development Mechanism (CDM); Joint Implementation (JI); and Emissions Trading. The CDM provides for Annex I Parties to implement projects that reduce emissions in non-Annex I Parties, or absorb carbon through afforestation or reforestation activities, in return for certified emission reductions (CERs, tCERs and lCERs) and assists the host Parties in achieving sustainable development and contributing to the ultimate objective of the Convention. Under JI, an Annex I Party (with a commitment inscribed in Annex B of the Kyoto Protocol) may implement an emission-reducing project or a project that enhances removals by sinks in the territory of another Annex I Party (with a commitment inscribed in Annex B of the Kyoto Protocol) and count the resulting emission reduction units (ERUs) towards meeting its own Kyoto target. Emissions Trading, on its side, provides for Annex I Parties to acquire units from other Annex I Parties. These units may be in the form of AAUs, removal units, ERUs, CERs, tCERs and lCERs.

Carbon transactions are defined as purchase contracts whereby one party pays another party in return for GHG emissions reductions or for the right to release a given amount of GHG emissions, which the buyer can use to meet its compliance—or corporate citizenship—objectives concerning climate change mitigation. Payment is made using one or more of the following forms: cash; equity; debt;

convertible debt or warrant; or in-kind contributions such as providing technologies to abate GHG emissions (UNFCCC 2008).

There is a range of active programs to manage GHG emissions that establishes a market by setting a target (absolute cap or intensity target) and that allow mandated participants to trade emissions allowances in order to meet compliance requirements at the lowest possible cost. Various emissions trading schemes exist inside and outside the scope of the Kyoto Protocol. These trading schemes are part of the commitment of States or companies to reduce their GHG emission. The EU Emissions Trading System (ETS) is still the most prominent carbon market.

With these mechanisms, Parties became able to access cost-effectively opportunities to reduce emissions or to remove carbon from the atmosphere, in other countries. If on the one hand the emission reduction cost depends on each region, on the other hand the effect for the atmosphere of limiting emissions is the same, regardless where the action is taken.

Although the second commitment period of Kyoto Protocol has begun on January 2013 (lasting until 2020), policy-makers have managed to agree on a new climate regime, in Paris, on December 12th 2015. The Paris Agreement seeks to accelerate the international response to the climate change challenge, by ensuring a global temperature rise, by 2100, below 2 °C above pre-industrial levels and to pursue the endeavor to limit the temperature growth to 1.5 °C (UNFCCC 2018a).

Bearing in mind the dynamics of climate change negotiations towards a post-Kyoto regime, Table 2.1 presents an overview on the major deliverables of climate change policies, over time.

In view of the Paris Agreement, OECD (2017) estimated a necessary investment of USD 6.9 trillion per year, in the next 15 years. Still this cost could be offset over time by fuel savings reaching USD 1.7 trillion per year by 2030 (OECD 2017).

From a regional perspective, Fig. 2.4 presents the carbon dioxide emissions breakdown per country, in 2017. Although China is the highest CO_2 emitter from a national perspective, the US is by far the largest polluter if one takes into account the emissions per capita (Figs. 2.5 and 2.6).

The European Union (EU) has been leading the climate agenda for the last decade. The Europe 2020 strategy for smart, sustainable and inclusive growth included five headline targets that set out where the EU should be in 2020. One of them relates to climate and energy. Member States have committed themselves to reduce GHG by 20%, increasing the share of renewables in the EU's energy mix to 20% and achieving the 20% energy efficiency target by 2020 (European Council 2014).

Furthermore, the European Commission (EC) presented in 2011 an "Analysis of options beyond 20% GHG emission reductions: Member State results" (European Commission 2011c). The EC's communication explored the options, and related costs and benefits, for moving towards a 30% reduction. The EC advocated that reaching the 20% GHG emissions reductions target and the 20% renewables target by 2020 has lower costs than originally foreseen. The Commission considered that the 20% emissions reduction target is less expensive than was assumed in 2008,

Table 2.1 Overview on major deliverables on climate change policies (UNFCCC 2018a)

Year	Deliverable
1979	The first World Climate Conference takes place
1988	The Intergovernmental Panel on Climate Change is set up
1990	IPCC's first assessment report is released. IPCC and second World Climate Conference call for a global treaty on climate change. United Nations General Assembly negotiations on a framework convention begin
1991	First meeting of the Intergovernmental Negotiating Committee takes place
1992	The Intergovernmental Negotiating Committee adopts UNFCCC text. At the Earth Summit in Rio, the UNFCCC is opened for signature along with its sister Rio Conventions, UNCBD and UNCCD
1994	UNFCCC enters into force
1995	The first Conference of the Parties (COP1) takes place in Berlin
1996	The UNFCCC Secretariat is set up to support action under the Convention
1997	Kyoto Protocol formally adopted in December at COP3
2001	Release of IPCC's Third Assessment Report. Bonn Agreements adopted, based on the Buenos Aires Plan of Action of 1998. Marrakesh Accords adopted at COP7, detailing rules for implementation of Kyoto Protocol, setting up new funding and planning instruments for adaptation, and establishing a technology transfer framework
2005	Entry into force of the Kyoto Protocol. The first Meeting of the Parties to the Kyoto Protocol (MOP 1) takes place in Montreal. In accordance with Kyoto Protocol requirements, Parties launched negotiations on the next phase of the Kyoto Protocol under the Ad Hoc Working Group on Further Commitments for Annex I Parties under the Kyoto Protocol (AWG-KP). What was to become the Nairobi Work Programme on Adaptation (it would receive its name in 2006, one year later) is accepted and agreed on
2007	IPCC's Fourth Assessment Report released. Climate science entered into popular consciousness. At COP13, Parties agreed on the Bali Road Map, which charted the way towards a post-2012 outcome in two work streams: the AWG-KP, and another under the Convention, known as the Ad Hoc Working Group on Long-Term Cooperative Action Under the Convention
2009	Copenhagen Accord drafted at COP15 in Copenhagen. This was taken note of by the COP. Countries later submitted emissions reductions pledges or mitigation action pledges, all non-binding
2010	Cancun Agreements drafted and largely accepted by the COP, at COP16
2011	The Durban Platform for Enhanced Action drafted and accepted by the COP, at COP17
2015	On December 12th 2015, 195 nations approved the Paris Agreement, which seeks to undertake climate change ambitious efforts to combat climate change and adapt to its effects, with enhanced support to assist developing countries to do so

which means that the 30% reduction scenario has turned out less expensive too (European Commission 2011c).

In a complementary way, the EC has proposed the Europe 2020 flagship initiative for a resource-efficient Europe and has drawn a set of long-run policies on transport, energy and climate change. The EC has sketched out the key elements

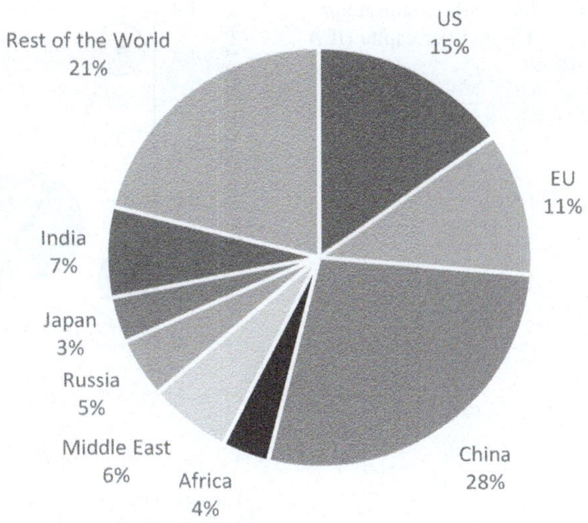

Fig. 2.4 CO_2 emissions breakdown per country [data from BP (2018)]

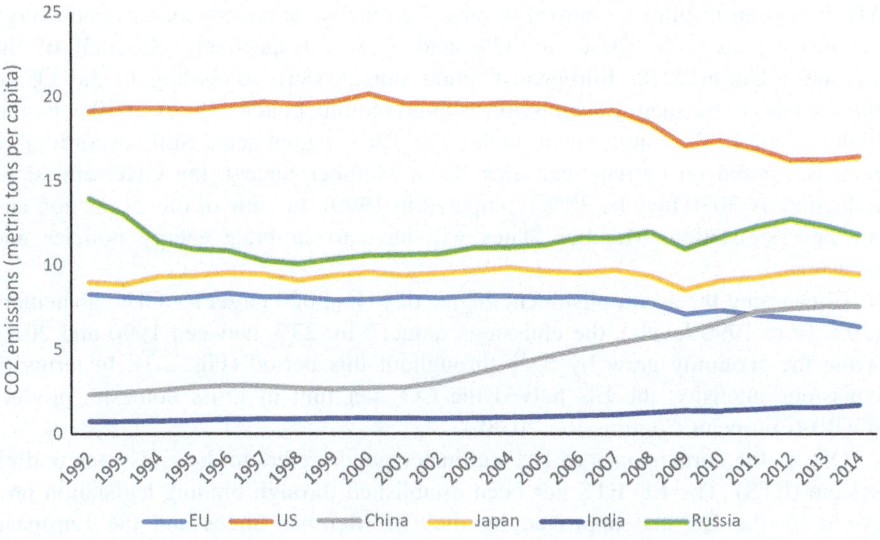

Fig. 2.5 Carbon emissions per capita [data from The World Bank (2018)]

that should shape the EU's climate action, road to a competitive low carbon economy by 2050 (European Commission 2011a). The European Council has also reconfirmed (in February 2011) the EU objective of reducing GHG emissions by 80–95%, by 2050, compared to 1990.

The EU has also adopted a climate and energy framework targeting the reduction of GHG by at least 40% by 2030 (compared to 1990) (European Council 2014).

Fig. 2.6 Carbon emissions per GDP and per capita (IEA 2017a)

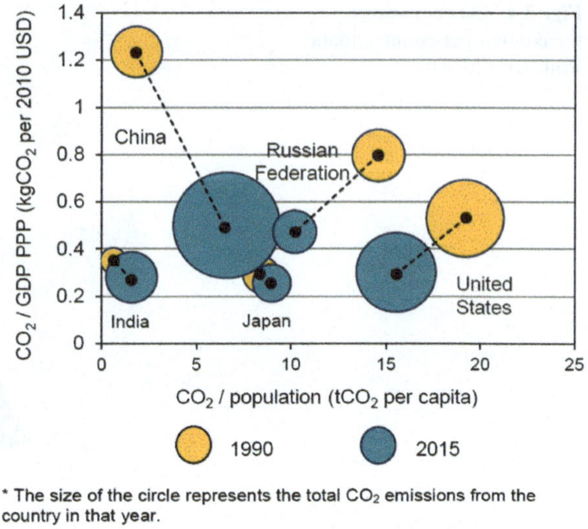

The European institutions agreed to raise the renewable energy sources and energy efficiency targets, by 2030, to 32% and 32.5%, respectively (Council of the European Union 2018; European Commission 2018a). According to the EC, if these goals are pursued, EU emissions should plummet circa 45% by 2030, which is higher than EU's commitment under the Paris Agreement. Still, according to forecasts based on current measures (from Member States), the GHG emissions reduction by 2030 may be 30% (compared to 1990). In view of this projection and the new legislation, Member States will have to calibrate energy policies and measures.

Concerning the accomplishment degree of EU's 2020 target for GHG abatement (20% from 1990 levels), the emissions reduced by 23% between 1990 and 2016, while the economy grew by 53% throughout this period (Fig. 2.7). In terms of emissions intensity, the EU halved the CO_2 per unit of gross domestic product (GDP) (European Commission 2018b).

One of the cornerstones of EU's climate agenda refers to the emissions trading system (ETS). The EU ETS has been established through binding legislation proposed by the EC and approved by the EU Member States and the European Parliament, being the first international trading system for CO_2 emissions in the world. Since 2007 the EU ETS applies not only to the 27 Member States, but also the other three members of the European Economic Area (Norway, Iceland and Liechtenstein) (European Commission 2008).

The EU ETS is a "cap and trade" system, that is to say it caps the overall level of emissions allowed but, within that limit, allows participants in the system to buy and sell allowances as they require. These allowances are the trading "currency" of the system. One allowance gives the holder the right to emit one tonne of CO_2. The cap on the total number of allowances is what actually creates scarcity in the market. At the end of each year, installations should surrender allowances

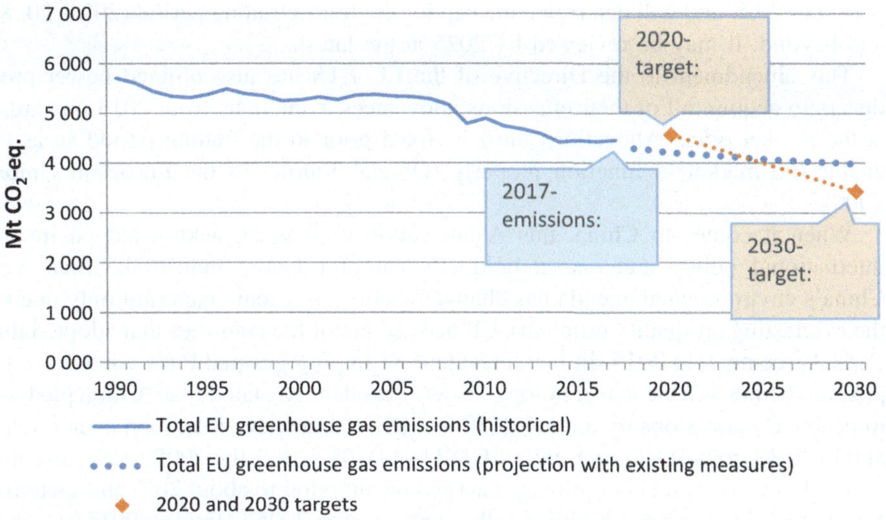

Fig. 2.7 Historical and projected GHG emissions, and abatement targets of the EU (European Commission 2018b)

equivalent to their emissions. Companies that keep their emissions below the level of their allowances can sell their excess allowances. Those facing difficulty in keeping their emissions in line with their allowances have a choice between taking measures to reduce their own emissions—such as investing in rather efficient technology or using less carbon-intensive energy sources—or buying the extra allowances they need in the market, or a combination of the two. Such choices are likely to be determined by relative costs. In this way, emissions are reduced wherever it is most cost-effective to do so (European Commission 2008).

For phase I (2005–2007) and phase II (2008–2012) of EU ETS, Member States drew up national allocation plans (NAPs), which determined their total level of emissions and how many emission allowances each installation in their country received. The European Commission presented in 2008 a proposal for amending the Directive 2003/87/EC (that establishes the EU ETS), which was published on June 5th 2009. The new allocation methodology in phase III brought bold differences, regarding the previous ones. Firstly, in phase III there is a Community-wide harmonized allocation method, in which auctioning is the basic principle for allocation, and from 2013 the total number of allowances decreases annually in a linear manner (European Commission 2011b). The starting point of this line is the average total quantity of allowances (phase II cap) issued by Member States for the 2008–12 period, adjusted to reflect the broadened scope of the system from 2013. The linear factor by which the annual amount decreases is 1.74% in relation to the phase II cap (Official Journal of the European Union 2009). The linear factor of 1.74% used to determine the phase III cap will continue to apply beyond the end of the trading period in 2020 and will determine the cap for the fourth trading period (2021–2028) and beyond. It may be reviewed by 2025 at the latest.

This amendment to the Directive of the EU ETS has also obliged power producers to acquire all of their emissions allowances at auctions from 2013 onwards, at the market price. Allocations must be fixed prior to the trading period so as to enable the market to function properly (Official Journal of the European Union 2009).

When it comes to China, this Asian nation used to be acknowledged in the international policy arena as a bold advocate for heavy industrialization. Yet China's environmental agenda has changed swiftly in recent years (not only due to the everlasting air quality problem). China was one of the countries that adopted the Paris Agreement, in 2015. In fact, president Xi Jinping grasped Paris conference to position China as a clean technology leader. President Xi stated that "China pledges to peak CO_2 emissions by around 2030 and strive to achieve it as soon as possible, and by 2030, reduce CO_2 per unit of GDP by 60–65% over the 2005 level, raise the share of non-fossil fuels in primary energy consumption to about 20% and increase forest stock by around 4.5 billion cubic meters over 2005" (Jinping 2015).

China has carried out a myriad of green policies, in order to curb greenhouse gases. It is on the way to decommission coal power plants and has launched, in 2017, a nationwide carbon emissions trading system in the power generation industry, gathering already roughly 1700 companies and aiming at broadening its scope to other industries. China has managed to slash 46% of CO_2 emissions per unit of GDP (from 2005 basis), achieving its goal to reduce carbon emissions by 40–45% from 2005 level by 2020 (UNFCCC 2018b). Still, China's greening could require \$6.4 trillion to \$19.4 trillion to finance the transition to a low carbon economy (Song 2018).

Concerning the US, despite the efforts of Clinton-Gore administration to approve Kyoto Protocol (rejected by the Senate and Congress), the Environmental Protection Agency (EPA) has only acknowledged the anthropogenic cause of climate change in 2009 (EPA 2009). That year President Obama pledged to curb US GHG emissions by 17%, from 2005 levels, by 2020. This policy was then put in place, in 2013, when The White House launched the Climate Action Plan, which enclosed three pillars: (i) cut carbon pollution in America; (ii) prepare the US for the impacts of climate change; and (iii) lead international efforts to combat global climate change and prepare for its impacts (The White House 2013).

One of the drivers of the Climate Action Plan consisted in the decarbonisation of the electricity sector. In view of that, a Clean Power Plan was issued in 2015, aiming at regulating carbon dioxide emissions from existing power plants and setting standards for new power plants (Harvard Law Review 2016). Eventually EPA's new administrator repealed the Clean Power Plan, in 2017, since this regulation has allegedly exceeded the agency's statutory authority (EPA 2017).

Regarding international climate policy, the Obama-Biden administration vowed, amongst the Paris Agreement, to reduce US GHG emissions from 26 to 28% by 2025 (from 2005 levels) (The New York Times 2017). However, after Donald Trump's election, the US decided to withdraw from the accord (Reuters 2017).

2.1.3 *Energy Trends*

Carbon Footprint

Sustainable energy systems are not only characterized by low carbon technologies or even demand side management. One should have in mind the myriad of United Nations sustainable development goals that ought to be pursued simultaneously. In view of that, energy sustainability—besides climate change mitigation—also encloses air quality and energy access objectives. Although the multiobjective challenge, there is an upside of delivering the sustainable development agenda at once. For instance, energy access can be tackled by supplying electricity based on renewable energy sources, which will also contribute for reducing the use of traditional biomass cook stoves. What's more, the adoption of the micro-grid concept in developing countries will attenuate the investment in bulk power generation and transmission infrastructure, hence limiting carbon emissions from fossil fuels and building a rather efficient energy system.

As referred in the current chapter, the Paris Agreement aims at peaking GHG emissions as soon as possible and then start a decreasing pace, road to net-zero emissions in the second half of the century. Before addressing the scenarios for sustainable energy systems, it is worth assessing current levels of carbon emissions.

Carbon emissions fall into four group of sectors: electricity and heat; industry (including oil and gas, steel and iron, cement and chemicals); consumer-related sectors (buildings, transport and waste); and land-use related sectors (forestry and agriculture). According to the IEA (2017a), the electricity and heat generation, and transport produced two-thirds of global CO_2 emissions in 2015, as shown in Fig. 2.8. These emissions are the result of anthropogenic activities, largely related with the use of fossil fuels (Fig. 2.9).

Taking into account global targets for emissions reductions and the world's actual carbon intensity, the IEA (2018a) drew a set of scenarios for long-term emission trends, as follows (Fig. 2.10). The Sustainable Development Scenario (SDS) sets a strategy in line with both the Paris Agreement and the United Nations Sustainable Development Goals. These commitments require a bolder effort from the power sector, through the exploitation of RES for universal energy access and maximizing the contribution of energy efficiency, in order to keep overall demand in 2040 at today's level (despite the electrification path of end-uses). Accordingly, the RES proportion in the energy mix would rise from 1/4 to 2/3 in 2040, whereas in the transport sector the renewables would rise from today's 3.5 to 19% in the target year (including electric mobility, charged by renewables-based electricity). Wind and solar photovoltaic would soon become the most competitive power sources in several geographies, totaling 1/3 of global electricity supplied in 2040. On the other hand, emissions reduction in transport, industry and buildings would be mostly achieved via energy efficiency—adopting efficient technologies and behaviors—and increasing the use of electricity as final energy.

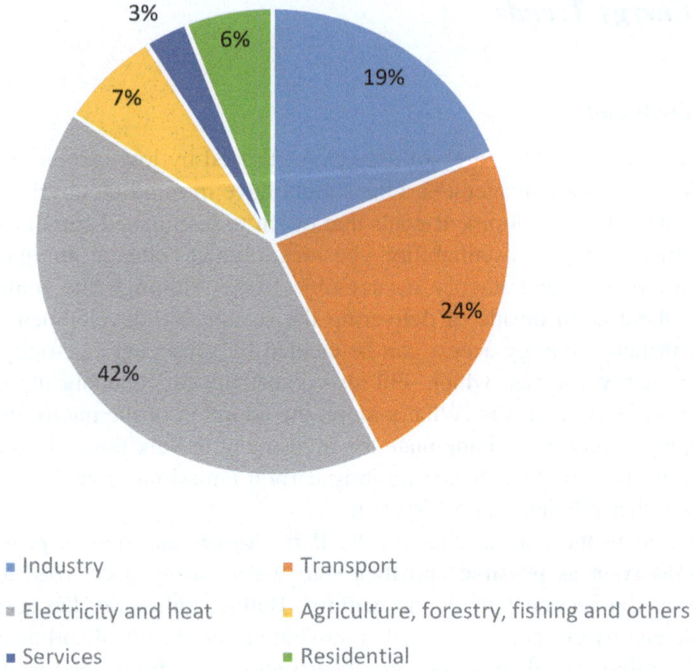

Fig. 2.8 World CO_2 emissions by sector in 2015 [adapted from IEA (2017a)]

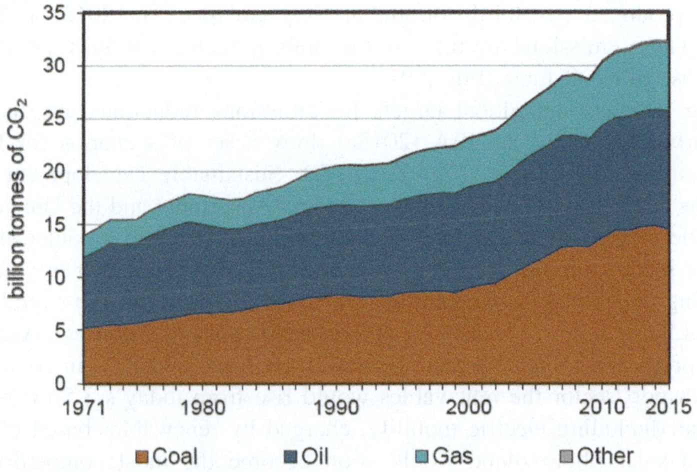

Fig. 2.9 CO_2 emissions by fuel in 2015 (IEA 2017a)

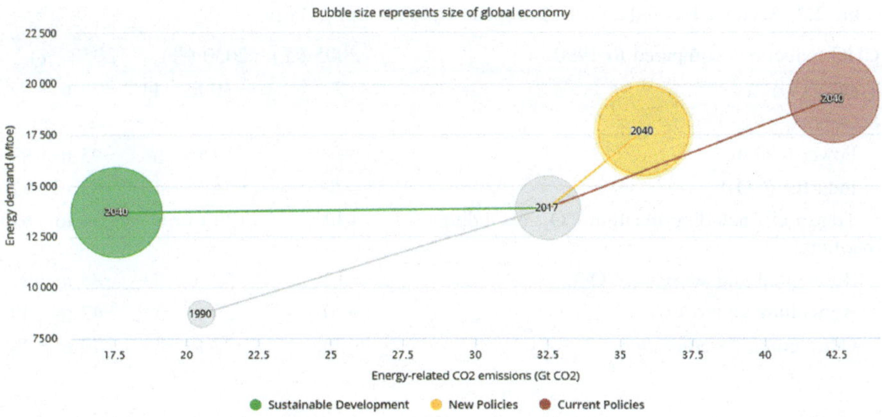

Fig. 2.10 World primary energy demand and energy-related CO_2 emissions by scenario (IEA 2018a)

According to the IEA, most energy related emissions are locked-in by 2040, especially the ones derived from coal-fired power plants (essentially early-life assets installed in Asia). Furthermore, current energy infrastructure (including the assets under construction) would count for 95% of the emissions allowed within international climate commitments. When comparing IEA's SDS with the New Policies Scenario (NPS), one concludes that the SDS would only require 15% of additional energy investment globally.

In the NPS energy demand rises more than 25% by 2040, taking into account higher income and world's population growth. The increase in energy demand is, however, smoothened due to the role of energy efficiency. In the supply side, RES and natural gas account for more than 80% of the energy demand.

Besides the IEA's scenarios, the European Commission (2011a) has drawn a sector analysis for a low carbon economy and Bloomberg New Energy Finance (2018a) estimated carbon emissions by fuel by 2050, as presented in Table 2.2 and Fig. 2.11, respectively.

Supply and Demand Outlook

Despite the broader political and social understanding of the environmental impact of fossil fuels, there is an asynchrony of the energy efficiency and decarbonization pace. According to the IEA (2018b), from 1971 to 2016 the world's total primary energy supply grew in a ratio of 2.5 and the fossil fuels are still the leading energy sources (Fig. 2.12).

Regarding energy demand, total final consumption has increased 2.25 times, from 1971. The world has observed different trends since 1971, which will be likely bolder in the next decades. On the one hand Asia's non-OECD region increased energy demand seven-fold, on the other hand OECD's demand has just grown 55% in the same period (IEA 2018b). According to the World Energy Outlook of IEA

Table 2.2 Sector GHG reductions (European Commission 2011a)

GHG reductions compared to 1990	2005 (%)	2030 (%)	2050 (%)
Total	−7	−40 to −44	−79 to −82
Sectors			
Power (CO_2)	−7	−54 to −68	−93 to −99
Industry (CO_2)	−20	−34 to −40	−83 to −87
Transport (including aviation CO_2, excluding maritime)	+30	+20 to −9	−54 to −67
Residential and services (CO_2)	−12	−37 to −53	−88 to −91
Agriculture (non-CO_2)	−20	−36 to −37	−42 to −49
Other non-CO_2 emissions	−30	−72 to −73	−70 to −78

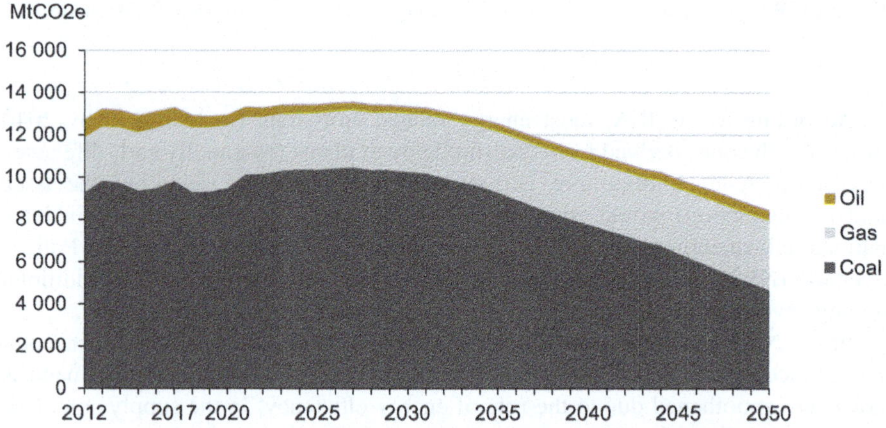

Fig. 2.11 Power sector CO_2 emissions by fuel by 2050 (Bloomberg New Energy Finance 2018a)

(2018a), under current and planned policies, energy demand is expected to rise about 25% by 2040. Additionally, in coming years the non-OECD region will be responsible for the bulk increase of energy demand, as shown in Figs. 2.13 and 2.14.

Statistics of IEA (2018a) also show the increasingly high concentration of energy demand, from a geographic perspective. In 2016, the ten countries with the largest energy demand accounted for 62% of global energy demand, whereas in 1971 this concentration was just 56%. Accordingly, changes in regional energy supply and demand profiles lead to meaningful shifts in global energy trade flows (Fig. 2.15).

From an energy per capita viewpoint, the US gathers nearly 5% of the world's population and consumed 16% of the energy, in 2016. The other way around, China and India depict lower energy per capita figures than the US. Indeed, in 2016 China

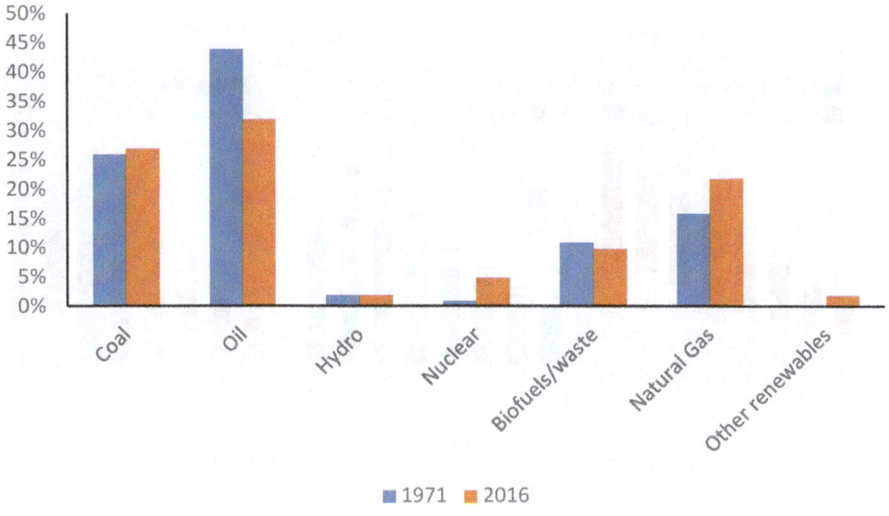

Fig. 2.12 Primary energy supply by source [data from IEA (2018b)]

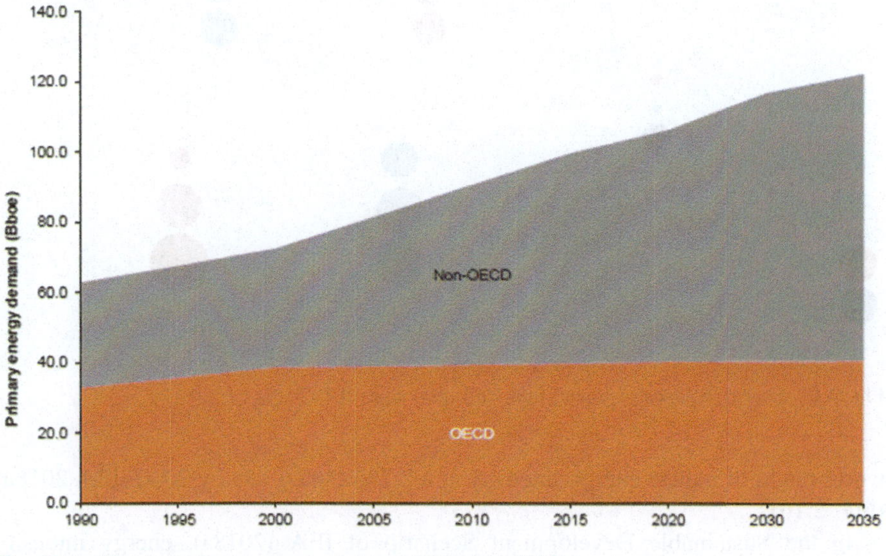

Fig. 2.13 Trend of primary energy demand (Citigroup 2013)

and India's share of global energy demand was 22% and 6%, respectively, and accounted for 20 and 19% of the population.

Besides energy per capita, one should also bear in mind the global energy intensity, which actually is one of the targets of UN's Agenda for Sustainable Development (as a metric for energy efficiency). To comply with UN's agenda, the

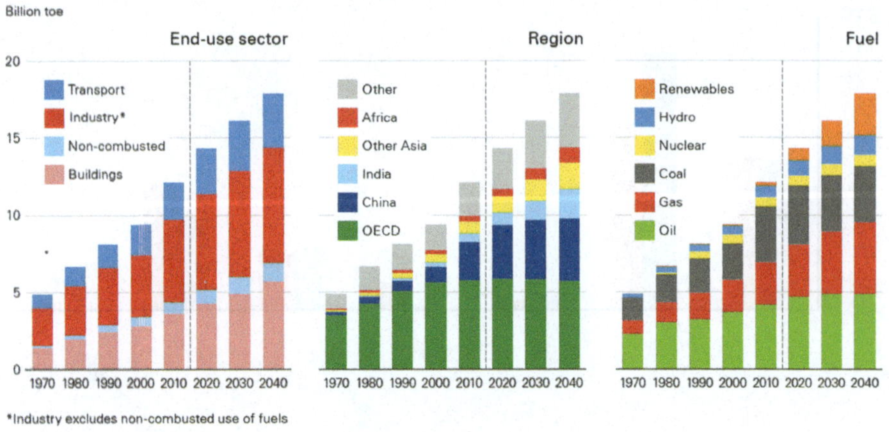

Fig. 2.14 Primary energy demand per sector, region and fuel (BP 2019)

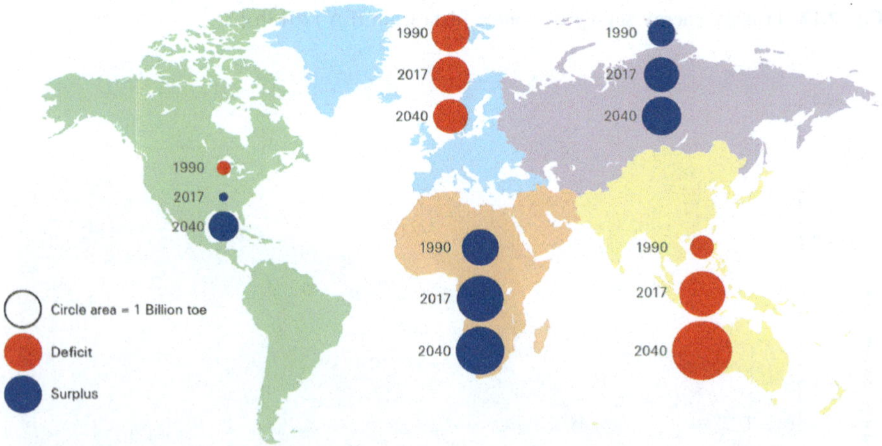

Fig. 2.15 Energy balance of traded fuels (oil, gas, coal) (BP 2019)

world ought to reduce energy intensity at a 2.7% annual rate by 2030 (IEA 2018a) (Fig. 2.16).

In the Sustainable Development Scenario of IEA (2018a), energy intensity decreases by 3.4%/year, whereas for the New Policies Scenario this indicator declines 2.3%/year from 2017 to 2040.

As stated by BP (2019) in the its energy outlook, primary energy demand trend depends of the world population evolution and higher living standards. Since global GDP more than doubles by 2040 and energy consumption increases by just a third, the energy demand growth is offset by the reduction in energy intensity (BP 2019) (Fig. 2.17).

Fig. 2.16 World's annual variation of energy intensity (IEA 2018a)

Fig. 2.17 Increase in primary energy demand (BP 2019)

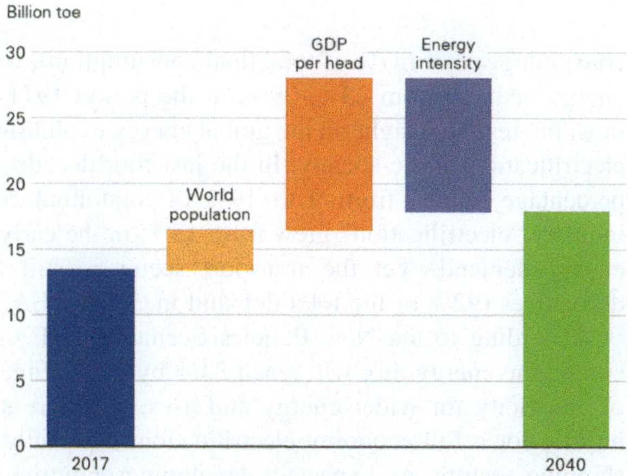

Yet the energy intensity indicator is not subject to a scientific consensus. On the one hand it is worth considering externalities of the GDP evolution in the energy demand, but on the other hand the energy intensity of a fast-growing region is benefited (in theory, even with a poor performance in energy use).

Additionally, energy demand is correlated with human progress, namely the United Nation's Human Development Index (HDI). Increases in energy consumption up to around 100 Gigajoules (GJ) per capita are linked with substantial improvements in human development and well-being. Roughly 80% of the population lives in countries where average energy demand is lower than 100 GJ/capita (BP 2019) (Fig. 2.18).

In view of the previous findings on global energy statistics, it is worth analyzing the energy demand by sector in the last decades. The industry has been the largest

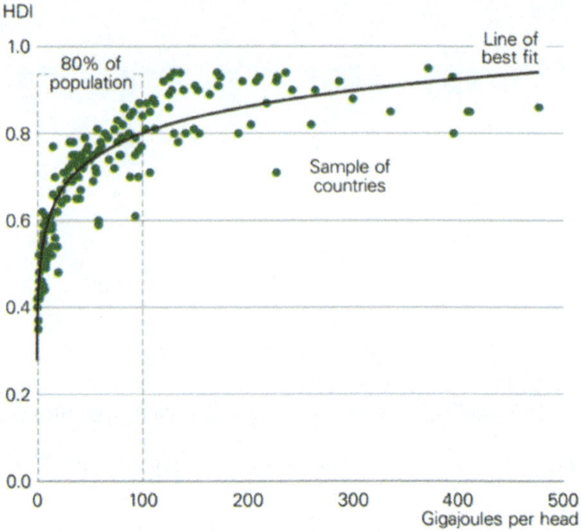

Fig. 2.18 Human development index and energy consumption per head, 2017 (BP 2019; UNDP 2019)

consuming sector (37% of total final consumption), while the transport has risen the energy demand from 23 to 29%, in the period 1971–2016 (IEA 2018b). Still, the most interesting insight on the global energy evolution refers to the higher degree of electrification of the society. In the last four decades the electricity use soared ten percentage points, from 9 to 19% of total final consumption. Specifically, the industry "electrification" grew from 15%, in the early 70^{ies}, to 27% of this sector's energy demand. Yet the transport sector is still almost totally fueled by oil derivatives (92% of the total demand in 2016) (IEA 2018c) (Figs. 2.19 and 2.20).

According to the New Policies Scenario of IEA (2018a), the electricity share among the energy mix will reach 24% by 2040 (Fig. 2.21). Although the potential of electricity for wider energy end-uses, there are still technical and/or financial hurdles for a full economy electrification, especially in the industry, aviation and shipping sectors. As expected, developing countries will grip the largest share of electricity demand growth, whereas developed economies will roughly reach a plateau in peak load and annual demand, due to energy efficient devices and rather educated demand-side behavior (Fig. 2.22). Although the upcoming momentum for electric mobility, this new power demand is not likely to be meaningful in developed countries. For instance, IEA (2018a) predicts that in a scenario of 100% of electric vehicles penetration (in vehicles sales), by 2040, the electricity demand growth would just increase to an average of 1.1%/year.

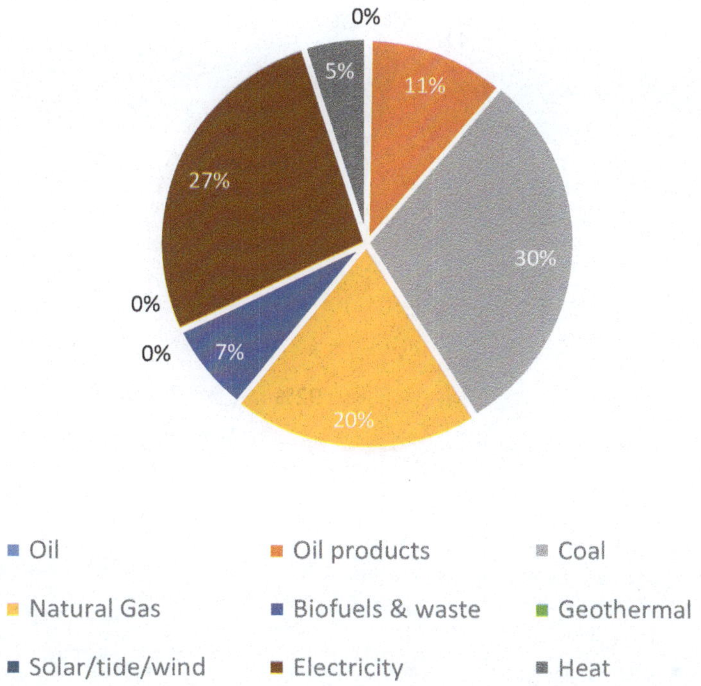

Fig. 2.19 Energy sources for the industry sector total final consumption—2016 [data from IEA (2018c)]

2.2 Power Systems

Sustainable and Digital Power Systems

As previously introduced in this chapter, the world is increasingly dependent of electricity for industrial processes, services and households needs (e.g. air and water heating) and mobility. If in the last century the electricity growth was vastly based in fossil fuel combustion (Fig. 2.23), one can predict with limited degree of uncertainty that the load increase in next hundred years will be mostly supplied by low carbon technologies (Figs. 2.24 and 2.25).

Besides the environmental driver—which presides today's energy policy—the future pace of electricity decarbonization will be determined by each technology cost, i.e. the multiyear trend for the levelized cost of energy (LCOE) per source. In fact, solar PV and onshore wind are reaching grid parity, whereas offshore wind, tidal and wave energy still require some type of support. The LCOE of onshore wind and solar PV projects, to be commissioned by 2023, is expected to range from 20 to 50 USD/MWh. That is to say, this means that these technologies will halve the contract price from 2017 to 2023 (IEA 2018d). Actually, solar PV is deemed to

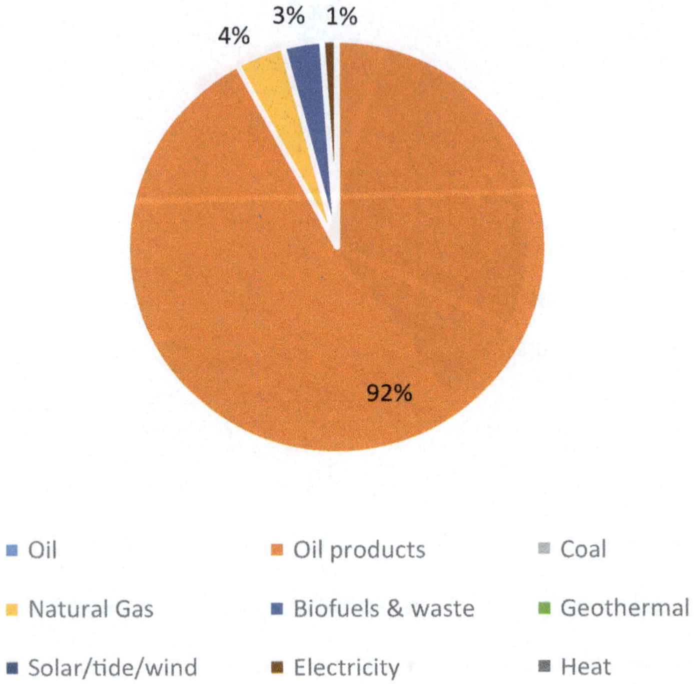

Fig. 2.20 Energy sources for the transport sector total final consumption—2016 [data from IEA (2018c)]

stand out as the most promising energy source, outranking wind power in the next few years, hydropower within 15 years and coal before 2040. In Portugal, a 2019 auctioning process for 1.4 GW of solar power has reached 20 €/MWh of bidding price (half of the Iberian wholesale price) (Expresso 2019). Accordingly, environment and market drivers might be responsible for a major electricity greening, moving from 2017's 25% of RES in the generation, to 40% by 2040 (IEA 2018a).

Although non-OECD countries are expected to grasp future global electricity rise, the upside comes from the way this demand will be supplied. Rather specifically, China and India are committed to tackle the load increase with RES, being accountable for over half of global solar PV capacity additions (IEA 2018a) (Fig. 2.26).

Yet the energy decarbonization trend brings a set of demanding challenges to the power system operation. There is a number of countries where RES provide a considerable share of electricity amongst the energy mix, e.g. Portugal, Spain and Denmark. In Portugal, the massive integration of RES is already on the way. Between March 9th and 12th 2018, Portugal's entire load was supplied by RES (mostly hydro and wind). This phenomenon underpinned the importance of flexibility to power grids, provided by peaking plants, interconnection capacity and

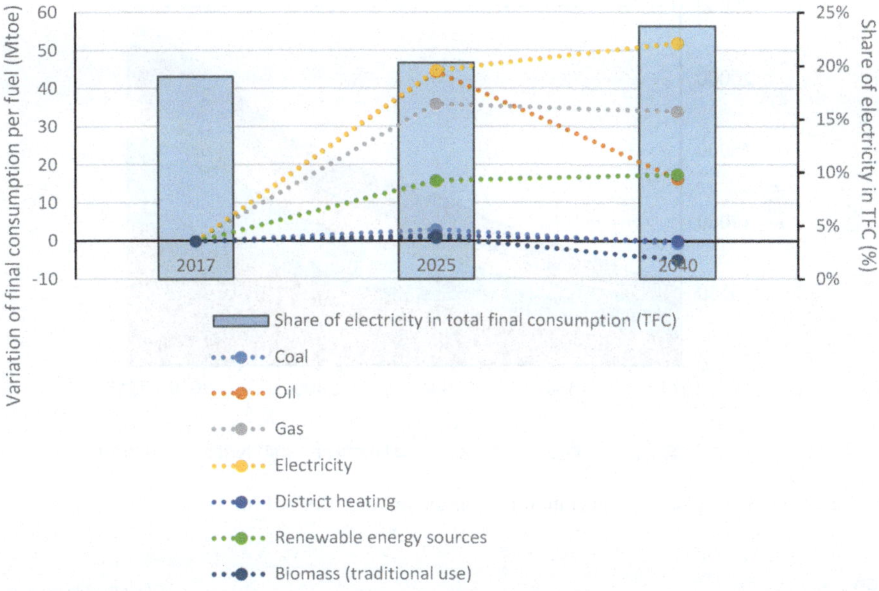

Fig. 2.21 Variation of total final consumption (TFC) by fuel and share of electricity in TFC, for IEA's New Policies Scenario [data from IEA (2018a)]

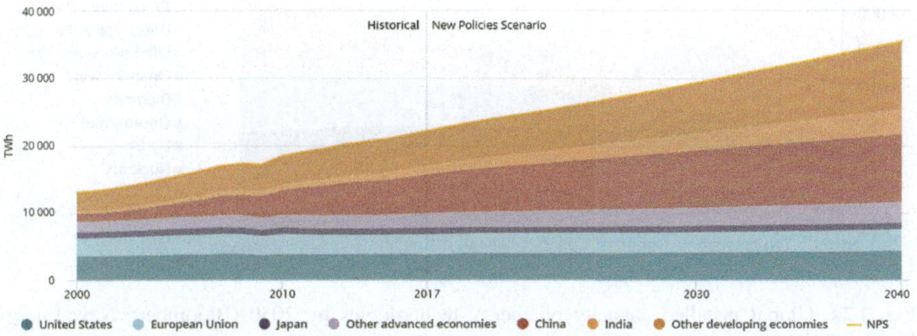

Fig. 2.22 Trend of global electricity demand by region (IEA 2018a)

energy storage. In this period of intense wind and hydro resources, Portugal's system stored (in pumped hydro facilities) 1463 MW (maximum) and exported to Spain up to 3967 MW (REN 2018) (Fig. 2.27).

Still in the Iberian Peninsula, Spain's power system has achieved a new wind generation record (43.2% of the electricity consumed, on January 23rd 2019), which together with other RES totaled 53.3% of the power supply. Currently, Spain gathers roughly 23 GW of installed capacity of wind power. Clean technology will

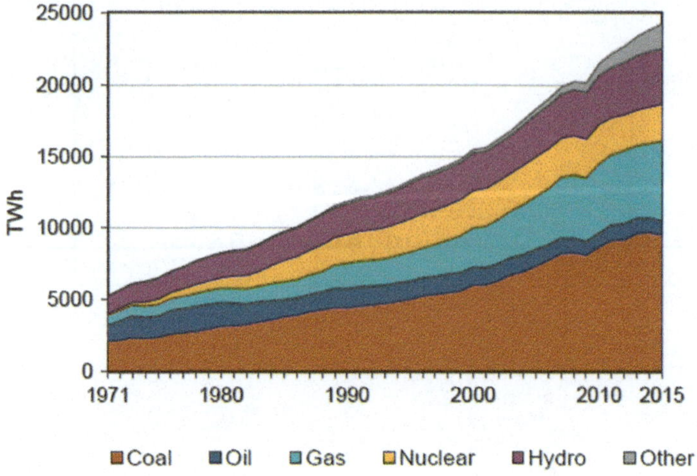

Fig. 2.23 Global electricity generation by energy source (IEA 2017a)

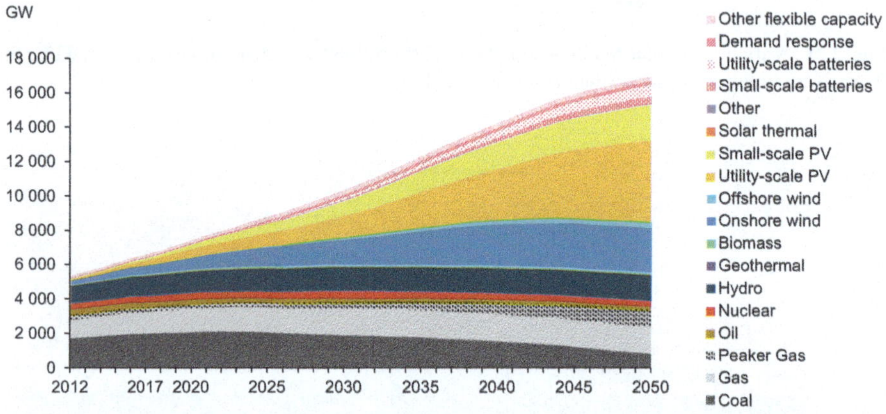

Fig. 2.24 Global installed capacity of energy technologies by 2050 (Bloomberg New Energy Finance 2018a)

have to grow 5 GW/year, by 2030, to fulfill the EU energy transition targets (REVE 2019).

In Denmark, wind power accounted for 43% of the electricity generated in 2017, totaling 14,700 GWh. By 2020, wind energy is expected to supply more than 50% of the electricity demand (Danks Energi 2018).

Besides these three examples of high amounts of RES in the generation mix, one should point out the bold renewable capacity growth in countries like China, US and Germany (Tables 2.3, 2.4 and 2.5).

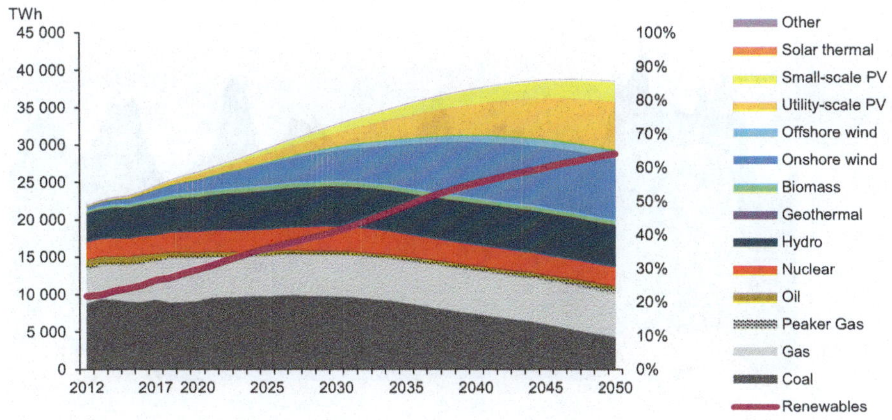

Fig. 2.25 Global power generation per technology by 2050 (Bloomberg New Energy Finance 2018a)

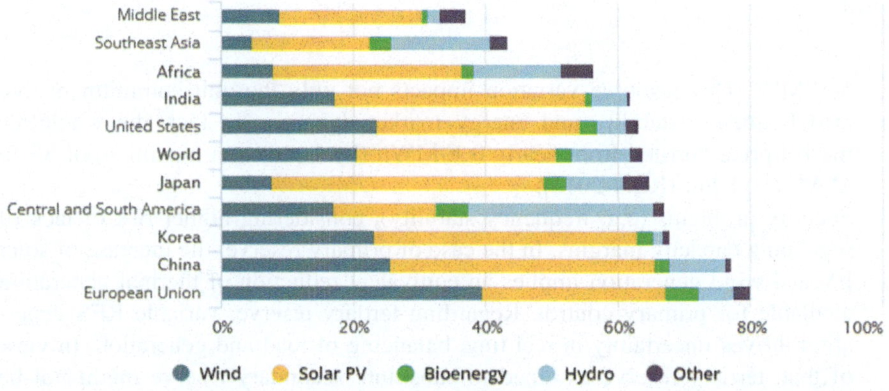

Fig. 2.26 Share of RES in total gross capacity additions per region in 2018–2040 (IEA 2018a)

This new energy *modus operandi* has been impacting the utilities' day-to-day operations, as listed next.

- Power Flow inversion in secondary substations, High Voltage/Medium Voltage substations and even in 150/60 kV substations. Indeed, Fig. 2.28 shows an example of power flow inversion, in a VHV/HV substation in Portugal, due to the injection of high amounts of variable RES in transmission grids.
- Increase of congestions on transmission and distribution grids at daily off-peak periods.
- Balancing of generation surplus in valley hours. For instance, on April 20th 2013 Portugal's power system has observed a sudden drop of wind generation: in just 8 h wind power slumped 80% of its global output, from roughly 2750 to

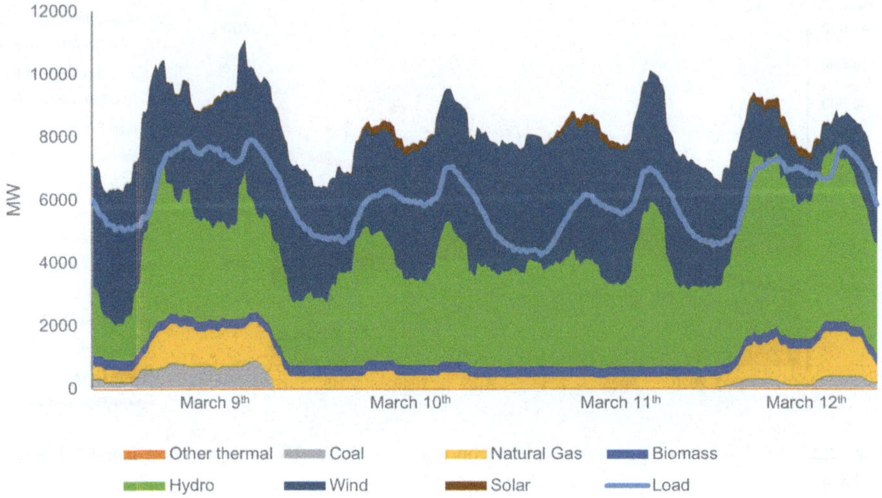

Fig. 2.27 Portugal's electricity supply and demand between March 9th and 12th 2018 (REN 2018)

550 MW. This resource variation impacts not only the unit commitment, but also frequency stability and energy market dynamic. In fact, the wholesale market price boosted from nearly 0 €/MWh (by 1 pm) to a maximum of 33 €/MWh at 11 pm (REN 2013).

- Security problems (e.g. frequency stability), considering higher risk of lack of regulating capacity margins. In the case of primary reserve, the increase of solar PV and wind generation implies an equivalent reduction of thermal generation available for primary control. Regarding tertiary reserve, variable RES generation thrives uncertainty in real time balancing of load and generation. In view of that, tertiary reserve assigned to substitute secondary reserve might not be always enough.

Besides these operational hurdles, system operators will have to manage an increasingly uncertain load profile, bearing in mind the advent of the smart grid concept. In truth, the smart grids were drawn to support the RES integration (besides the obvious customer data enrichment and the utility's cost optimization). However, the smart grid itself will lead to a new and major role of loads in the system operation (becoming controllable and price elastic), taking into account the empowerment of the consumer in the electricity value chain. Indeed, the smart grid has been enabling the transformation of the electricity value chain, from a one-way vertical process to a dynamic customer-centric system.

In addition, the power system is foreseen to broaden its boundaries to new services, following public policies towards the holistic integration of energy and transport objectives. The electric mobility path will introduce major challenges to the distribution system operator (DSO) but also to the transmission system operator

Table 2.3 Wind onshore capacity per region in 2018 [data from GWEC (2019)]

Region breakdown	New wind onshore capacity[a] (MW) 2018	Total wind onshore capacity[b] (MW) 2018
Total	**46,820**	**568,409**
Americas	**11,940**	**135,041**
USA	7588	96,635
Canada	566	12,816
Brazil	1939	14,707
Mexico	929	4935
Argentina	494	722
Chile	204	1621
Other Americas	220	3605
Africa, Middle East	**962**	**5720**
Egypt	380	1190
Kenya	310	336
South Africa	0	2085
Other Africa	272	2109
Asia-Pacific	**24,902**	**256,320**
China	21,200	206,804
India	2191	35,129
Australia	549	5362
Pakistan	400	1189
Japan	262	3661
South Korea	127	1229
Vietnam	32	228
Philippines	0	427
Thailand	0	648
Other Asia	141	1643
Europe	**9016**	**171,328**
Germany	2402	53,180
France	1563	15,307
Sweden	717	7216
United Kingdom	589	13,001
Turkey	497	7370
Other Europe	3248	75,435

[a]Gross capacity
[b]Net capacity adjusted by decommissioning

(TSO). Besides the Electric Vehicles (EVs) effect on the daily load profile and wholesale prices, electric mobility may bring a set of ancillary services within the scope of the system operators, in view of the Vehicle-to-Grid (V2G) approach. In view of that, a growing interplay between TSOs and DSOs is deemed crucial.

Table 2.4 Wind offshore capacity per region in 2018 [data from GWEC (2019)]

Region breakdown	New wind offshore capacity[a] (MW) 2018	Total wind offshore capacity[b] (MW) 2018
Total	**4496**	**23,140**
Europe	**2661**	**18,278**
United Kingdom	1312	7963
Germany	969	6380
Belgium	309	1186
Denmark	61	1329
Netherlands	0	1118
Other Europe	0	302
Asia-Pacific	**1835**	**4832**
China	1800	4588
South Korea	35	73
Other Asia	0	171
Americas	**0**	**30**
USA	0	30

[a]Gross capacity
[b]Net capacity adjusted by decommissioning

Table 2.5 Solar capacity per region in 2018 [data from IRENA (2019)]

Region breakdown	New solar capacity (MW) 2018	Total solar capacity (MW) 2018
World	**94,763**	**485,826**
Africa	**1809**	**6093**
Asia	**64,076**	**274,866**
China	44,216	175,032
India	9225	27,098
Japan	6460	55,500
Europe	**9255**	**121,692**
Germany	3593	45,932
Italy	438	20,126
UK	332	13,108
North America	**10,540**	**57,118**
USA	8419	51,450
South America	**2052**	**5469**

The electric mobility is, nevertheless, a single piece amongst a wider smart grid infrastructure. Advanced metering is indeed the cornerstone of the smart grid platform, together with substation automation, power electronics, high voltage direct current (HVDC), asset sensing and battery energy storage. This initial layer of the smart grid enables the integration of innovative technologies and processes,

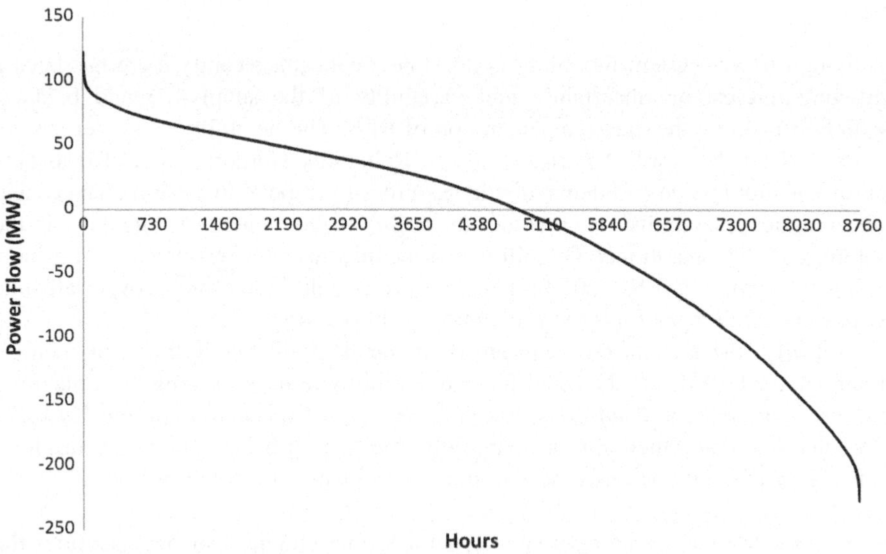

Fig. 2.28 Power flow inversion in a transmission grid substation, in Portugal

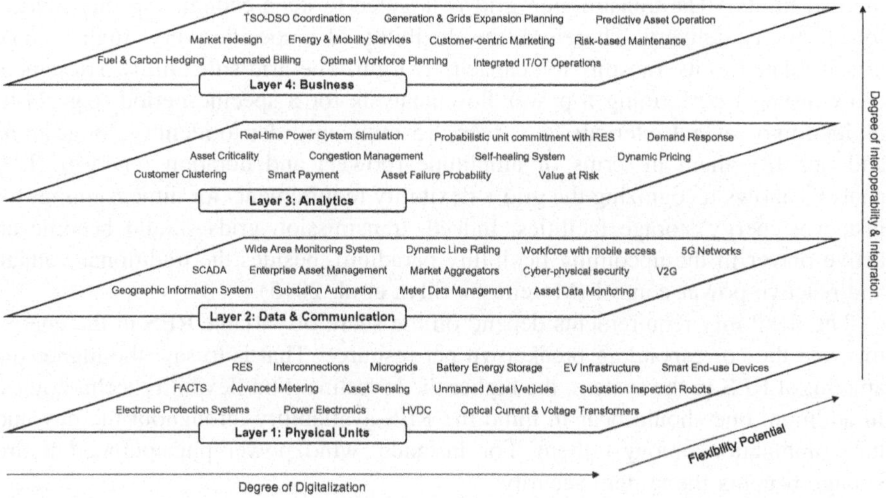

Fig. 2.29 Architecture of the Smart Grid

such as renewable energy self-consumption, demand response, dynamic pricing, V2G, predictive asset operation and optimal workforce planning (Fig. 2.29).

The upcoming smart power system will then present various challenges for both the supply and demand sides. Besides integrating large amounts of RES, the future power systems will have to ensure flexibility, stability and reliability.

Flexibility

Although power system flexibility is not a new concept, recently there has been a growing interest on uncertainty and variability of the supply-demand balance, mainly caused by the massive penetration of RES. The flexibility of power system is defined by the North American Electric Reliability Corporation (2010) as "the ability of supply-side and demand-side resources to respond to system changes and uncertainties. Flexibility also includes the ability to store energy for delivery in the future and the operational flexibility to schedule/dispatch resources in the most efficient manner" (NERC 2010). This is a broad definition which covers all time scales and challenges for system planning and operation.

Actually, there is lack of consensus in the flexibility definition and metrics. Lannoye et al. (2012a) advocated for two flexibility indices for long-term planning purposes, namely the *Insufficient Ramping Resource Expectation* and the *Period of Flexibility Deficit* which aim at quantifying the overall balance between the flexibility available and the one required, in a probabilistic and deterministic approach, respectively (Lannoye et al. 2012a, b).

Besides the supply and demand dynamics, one should also bear in mind the transmission and distribution grids role towards flexibility. The transmission grid impacts the flexibility of the power system, taking into account its typology and technical characteristics (e.g. power transformers capacity and overhead lines thermal limits). The transmission grid is targeted to keep enough capacity margin for electricity delivery. Therefore, the flexibility of a specific (very) high voltage grid is related to its transmission capacity margin. The grid's flexibility assessment can consist in performing a power flow analysis for a specific period (e.g. 24 h) divided into several intervals (e.g. every 15 min), in order to identify congestions and quantify them in terms of amplitude (power) and duration (energy). This analysis allows recognizing the grid's flexibility requirements for dimensioning, for instance, energy storage facilities. Indeed, transmission grids should become an active player in the upcoming flexibility paradigm, besides the traditional voltage and reactive power control (Moreira da Silva et al. 2015).

The flexibility requirements depend on the share of variable RES in the energy mix and the corresponding breakdown per resource. That is to say, the higher the amount of RES in the system, the higher the investment on flexibility technologies. In addition, one should bear in mind the RES availability throughout the day and the combination amongst them. For instance, wind power plus pumped hydro storage benefits the system security.

In the absence of a "silver bullet" to utterly tackle the operational problems raised by the variability of RES, energy planners should design blended strategies to strengthen the power system flexibility, taking into account the next possible options.

- "Peaking plants" (e.g. hydro units or gas turbines) which are able to follow load variations rapidly, due to their technical features (e.g. load gradients, power ramps, start-up times). Appropriate price signals would attract investments in these "conventional" generation units, ensuring they provide flexible "back-up" capacity.

- Active demand side management, in which suppliers could stimulate consumers to shift their consumption from "peak" to "off-peak" hours, by rewarding them with economic incentives.
- Transmission and distribution infrastructure expansion, reinforcement and automation to mitigate grid congestions. The development of smart grids would allow the system to effectively cope with bidirectional flows of electricity, related to the phenomenon of decentralized generation and "prosumers", and manage overload problems.
- Development of cross-border transmission capacity to strengthen the system's flexibility. For instance, a country could benefit from peaking units available in foreign countries, when load is increasing, and export electricity at times of high variable RES and low demand.
- Implementation of electricity storage, which is seen as an inevitable strategy to cope with larger variability and intermittency of supply. Electricity storage should be considered as one of the many means to provide various services to the system, such as capacity firming, capacity accommodation, voltage and frequency regulation or back-up capacity.
- Design of charging strategies of EVs, including V2G. EVs could be called by the system operator to provide ancillary services, through controllable battery charging and even power supply, without jeopardizing the mobility purpose.

In view of the aforementioned options to handle massive RES penetration, short-span flexibility technologies (e.g. batteries, demand response and V2G) could ease the introduction of 16% more wind and solar power in Europe, by 2050 (Fig. 2.30).

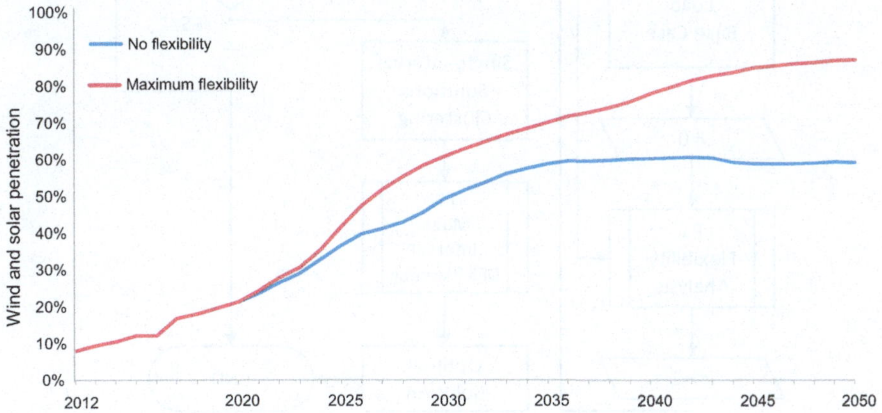

Fig. 2.30 Contribution of short-span flexibility for wind and solar penetration in Europe by 2050 (Bloomberg New Energy Finance 2018a)

Recent research advocates for the commissioning of grid-scale batteries, within the TSO's asset portfolio, for specific flexibility needs (without jeopardizing market operation). The use of energy storage brings additional flexibility to the TSO (in terms of congestion management) than the conventional options, namely capacity expansion. That is to say, some investments in grid expansion could be postponed by the adoption of energy storage units. Consequently, the system costs for grid development could be curbed and the assets optimized.

Indeed, the problem of planning dispersed energy storage (DES) in transmission grids is gaining *momentum* among the TSOs. While most of the literature tend to investigate energy storage within the scope of RES integration, energy trading and ancillary services supply, DES could also enhance grid flexibility due to its fast response and modularity.

In Moreira da Silva et al. (2015), an optimization algorithm was presented for planning DES in transmission grids. The algorithm gathers two main parts: (P1) Flexibility Analysis; and (P2) Multi-Interval DES Planning. The objective of the algorithm is to find the combination of site and size of DES facilities, in the grid, that minimizes the solution cost. This means finding the optimal power and energy requirements for DES infrastructure (Fig. 2.31).

The first part of the algorithm simulates the grid for each of the considered intervals. If grid congestion is detected, a simulation of the single-interval Evolutionary Particle Swarm Optimization (EPSO) algorithm—provided in module P1.1—is performed to solve the congestion using DES (Fig. 2.32).

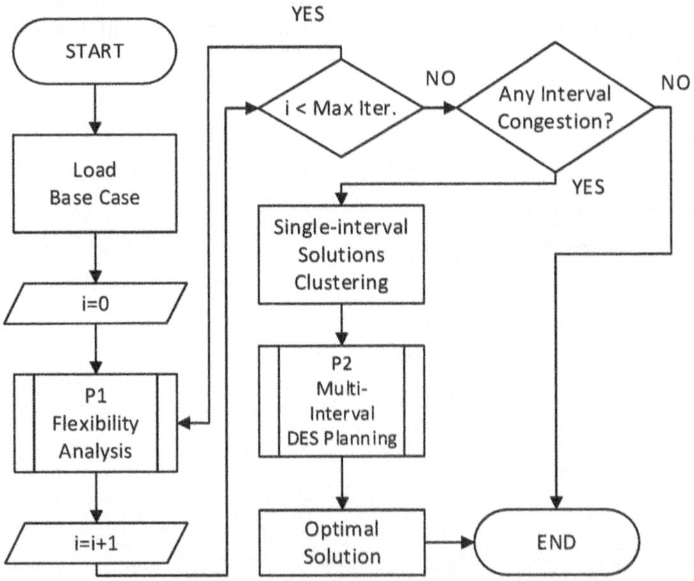

Fig. 2.31 Structure of the algorithm for planning dispersed energy storage (Moreira da Silva et al. 2015)

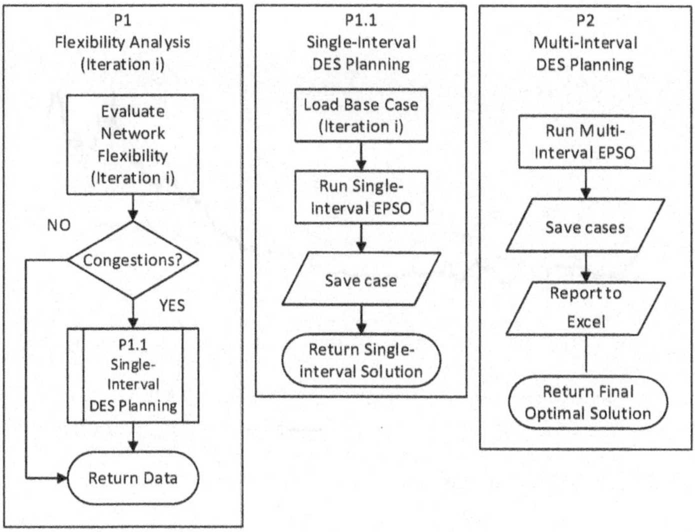

Fig. 2.32 Logic of the EPSO algorithm for DES planning (Moreira da Silva et al. 2015)

After the whole period simulation, the algorithm solves the multi-interval DES planning problem by finding the optimum set of site and size for DES, that minimizes the power and energy requirements of the DES system for the full period. In order to keep the previously simulated information, ignoring temporal dependence, the obtained solution is included in the original swarm of the multi-interval EPSO, as an approximation of the optimal solution. The algorithm was tested in a theoretical power system and provided solutions that minimize the total storage capacity and energy requirement, while reducing the investment cost.

2.3 Natural Gas

The Lever for a Clean and Secure Energy System

Natural Gas has gained *momentum* in the last decade, fueled by energy decarbonization pledges—that eventually turned out into legally binding targets in some regions, as in the EU—coupled with carbon pricing mechanisms (e.g. EU's Emissions Trading System). With carbon cap-and-trade, natural gas could outrank coal in the power generation merit order, if CO_2 price assumes environmentally integrity. If one performs a simple sensitivity analysis of the CO_2 price impact on the electricity market, gas future depends highly on the environmental integrity of the carbon price. According to RTE (2016), for 0 €/tonCO_2 coal generation is traded at 25 €/tonCO_2, whereas natural gas units are dispatched at 40 €/tonCO_2.

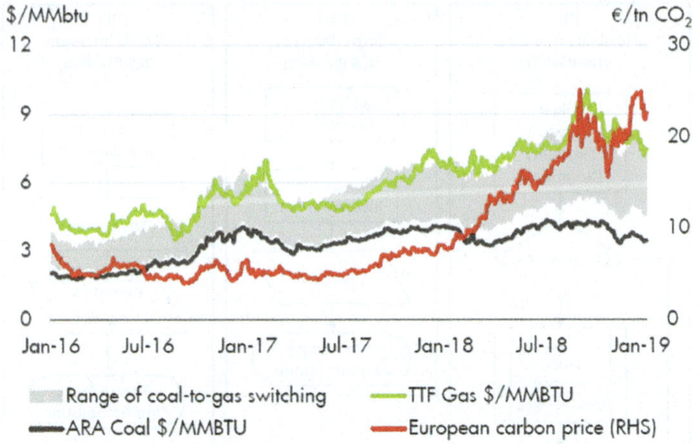

Fig. 2.33 Coal-to-gas shifting in European power sector (Royal Dutch Shell 2019)

For the scenario of current carbon price, namely 22 €/tonCO_2 (in March 2019), the most efficient gas units can be competitive with the least efficient coal power plants. If one analyses just the set of the most efficient gas and coal power plants, gas only outranks coal, in the economic merit order, when carbon reaches 50 €/tonCO_2 (RTE 2016). This insight is corroborated by the graph presented in Fig. 2.33, in which coal-to-gas shift increases with the carbon price.

Several entities, such as Carbon Tracker (2018), believe the carbon price is likely to hike amongst the EU Emissions Trading System, as a reflex of the Paris Agreement targets (55 €/tonCO_2 by 2030).

Although the political will to reach carbon neutrality, the power generation mix ought to count on thermal units, in order to ensure system adequacy, security of supply and service quality. Taking into account the carbon emission factors of fossil fuel technologies, natural gas power plants come out as the most energy efficient solution to tackle fuel shifting (from coal), seasonality and ancillary services (Fig. 2.34).

The energy planning endeavor should, therefore, consider natural gas facilities (e.g. combined-cycle gas turbines) to partner with ambitious RES investment.

If the gas business case seems clear for the power generation market, the same cannot be said for the domestic segment. In the last mile of energy systems, there is a competition between electricity and natural gas, for heating purposes. In the advent of energy liberalization, retailers have become multi-utility suppliers, which has benefited electricity incumbents (traditionally, with a broader network than gas distribution concessions). On the other hand, since electricity is understood as a cleaner and safer energy vector, natural gas could be impacted by green taxes that jeopardize its competitiveness.

Yet the subjectivity should be replaced by thorough reasoning. A life-cycle analysis must be performed for every energy mix. For instance, if a given power

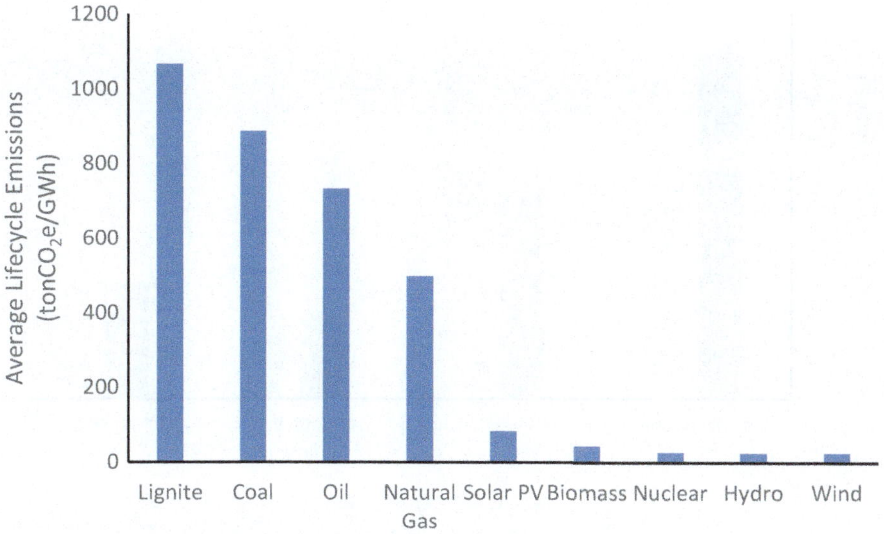

Fig. 2.34 Average Lifecycle Emissions for Power generation [data from World Nuclear Association (2011)]

generation mix is largely dominated by gas and coal units, natural gas becomes clearly more energy and environment efficient than electricity, for water and air heating. In the hypothetical case of a full renewable-based energy mix, electricity dominates natural gas as far as CO_2 is concerned, but still raises doubts in terms of energy efficiency (considering power generation, transmission and distribution losses) and cost-effectiveness. The electricity-gas competition for heating arises mostly in developed countries, whereas in developing nations one should add coal to the equation (with dramatic environmental externalities). Natural gas is also expected to raise its penetration in the industry and transport sectors, bearing in mind the pledges for plunging the use of emissions-intensive fossil fuels. For instance, industry will be the main driver for natural gas demand rise in China. From a worldwide perspective, industry will outplace power generation as the largest sector in gas demand growth. Natural gas will likely expand its use, including fertilizers in emerging economies and feedstock for petrochemicals in regions rich in natural gas (IEA 2018a) (Fig. 2.35).

Given the aforementioned context, natural gas will likely hold an important role not only for the power generation decarbonization (although increasingly limited to peaking and backup services), but also for transport, industry and residential energy demand. According to IEA (2017b), natural gas is the largest single fuel in the global mix in the Sustainable Development Scenario. Demand for natural gas is expected to increase by almost 20% by 2030, reaching a plateau at this level throughout the following decade (around 4.2 trillion cubic meters) (IEA 2017b).

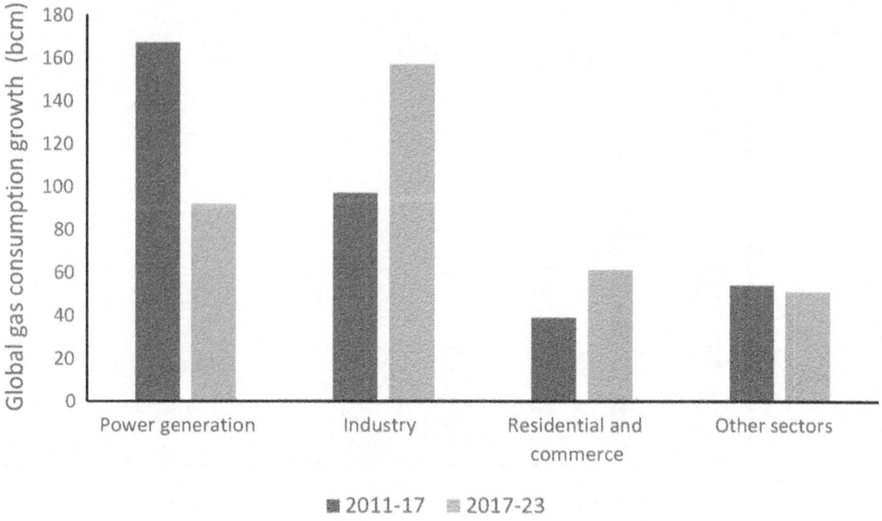

Fig. 2.35 Global gas consumption growth per sector [data from IEA (2018)]

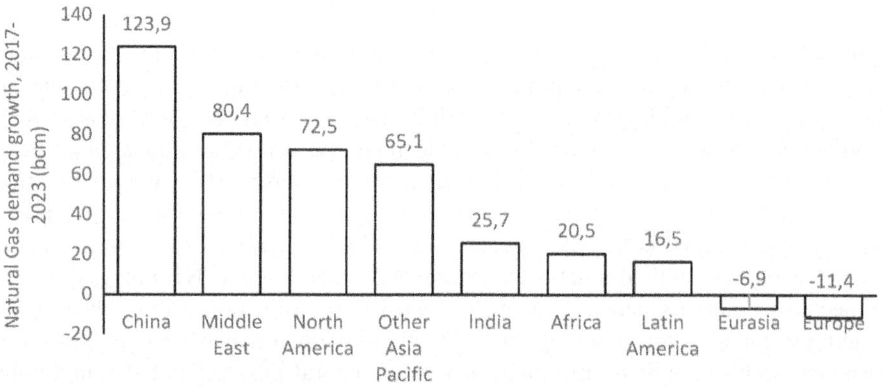

Fig. 2.36 Natural gas demand growth, 2017–2023 [data from IEA (2018)]

If on the one hand natural gas demand is constrained by the investment in RES and energy efficiency in developed countries, on the other hand Asian developing economies will drive gas demand growth by 2040 (IEA 2018a). China, in particular, will be accountable for 1/3 of global demand rise by 2022, taking into account its policy for air quality enhancement (Fig. 2.36).

Game Changers: Interconnections and LNG

In addition to the climate change agenda, energy dependence is at the core of national and international politics. Besides oil (out of the scope of this book), natural gas is subject to geopolitical strains and price volatility. Europe, in this context, is particularly exposed to Atlantic and Eastern conflicting agendas. Six EU member states depend on Russia as single gas supplier, and this energy vector provides over 1/4 of total energy demand in three of those countries (European Commission 2015). Additionally, 14 countries obtain more than 50% of their gas from Russia (Bloomberg 2019). As a result, EU's dependence of Russian gas triggered a political reaction by the European Commission, with the presentation and approval of the Energy Union package (Figs. 2.37 and 2.38).

The Energy Union, from a natural gas perspective, seeks to enhance system security and diversify energy supply. This strategy is being delivered through two main levers: (i) gas interconnections (within the EU and with Central Asia); and (ii) LNG terminals (to broaden the gas supply portfolio) (Fig. 2.39).

The European Commission aims at leveraging Central Asian gas supply, through the Southern Gas Corridor. On the other hand, Gazprom is leading the development of the Nord Stream 2 project, to link the European gas network (at Germany's Baltic coast) to Russian fields. This undersea pipeline will double the capacity of current infrastructure (the original Nord Stream) (Bloomberg 2019) (Fig. 2.40).

Concerning the second lever—LNG terminals—the EU advocates for the establishment of liquid gas hubs, with multiple suppliers, in Northern, Central, Eastern and Mediterranean areas, in order to speed up the security of supply goal. LNG is envisaged, by the EU, as a means to handle gas supply shortages from the transmission pipelines (i.e. acting as backup) (European Commission 2015). LNG brings flexibility to gas systems, by meeting seasonal consumption and hedging demand spikes (Fig. 2.41).

Complementarily, LNG investment soar could reduce the gap in the gas price traded worldwide. Yet so far LNG prices have been recurrently higher than pipeline gas, owing to the Asian bold demand and expensive costs for liquefaction, regasification and transport (Fig. 2.42). The European Union is tackling this hurdle by investing in transport infrastructure (linking LNG access points to the internal

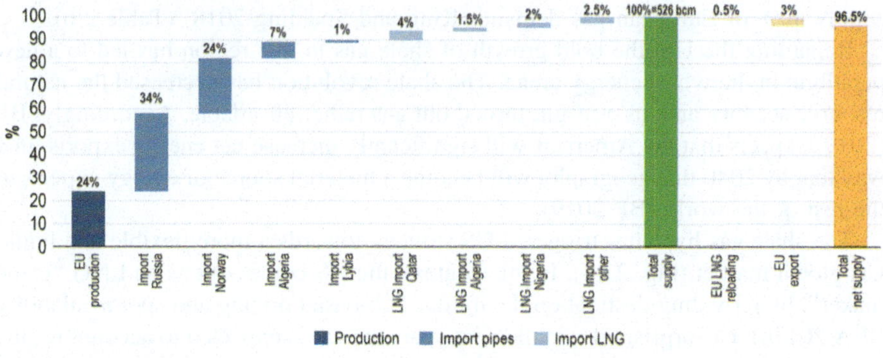

Fig. 2.37 EU gas supply portfolio by origin—2017 (100 = 526 bcm, %) (ACER 2017)

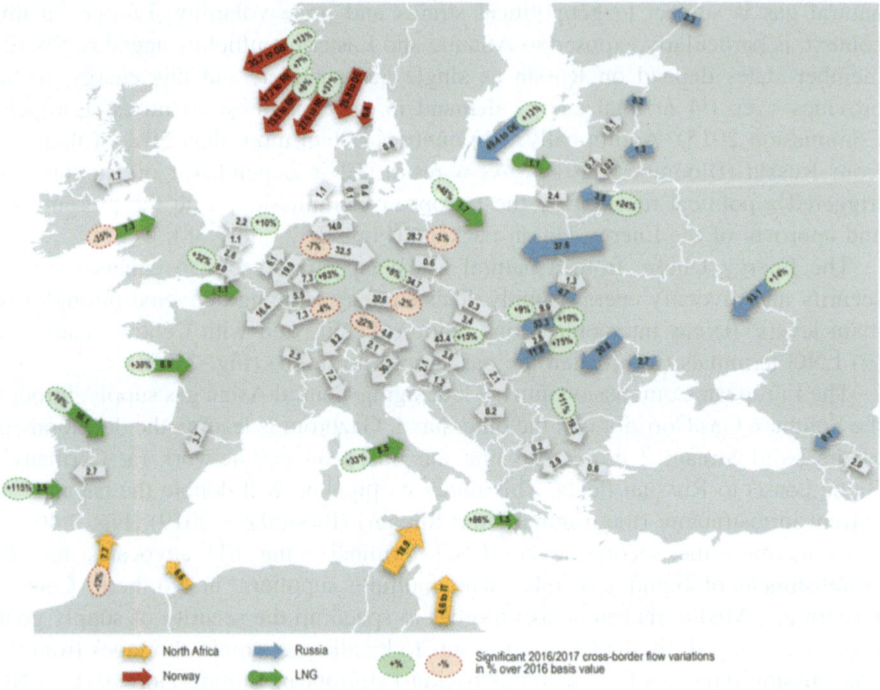

Fig. 2.38 EU Gas flow in 2017 (ACER 2017)

market) and withdrawing obstacles to LNG imports from the US and other LNG sources (European Commission 2015).

At this stage, one should highlight that except the cases of Norway and Russia (non-EU), Europe is fully equipped with LNG import terminals. Europe gathers 28 large-scale LNG import facilities (24 of which are in EU member states and 4 in Turkey) and 8 small-scale units. Europe's total regasification capacity from large-scale LNG terminals—nearly 230 billion cubic meters (bcm)—is enough to supply 40% of European gas demand (King and Spalding 2018) (Table 2.6).

Regarding the US, the bold growth of shale gas in this region has led to a new paradigm in the world energy arena. The shale revolution has decreased the natural resource scarcity and, as a result, turned out gas rather affordable. Accordingly, BP (2019) expects that the Americas will significantly increase net energy exports, in a way that by 2040 this geography will become a material source of energy exports to the rest of the world (BP 2019).

The shale gas hype has triggered US strategy towards a more flexible and liquid gas global market (Fig. 2.43). In this context, the US comes out as an LNG "game maker", by providing destination flexibility, hub-based pricing and spot availability (IEA 2017b). Unsurprisingly, in the next years the US is expected to account for the largest share of supply growth (with the highest export rise), whereas Europe

Fig. 2.39 EU projects of common interest for gas pipeline and LNG terminals (EC 2017)

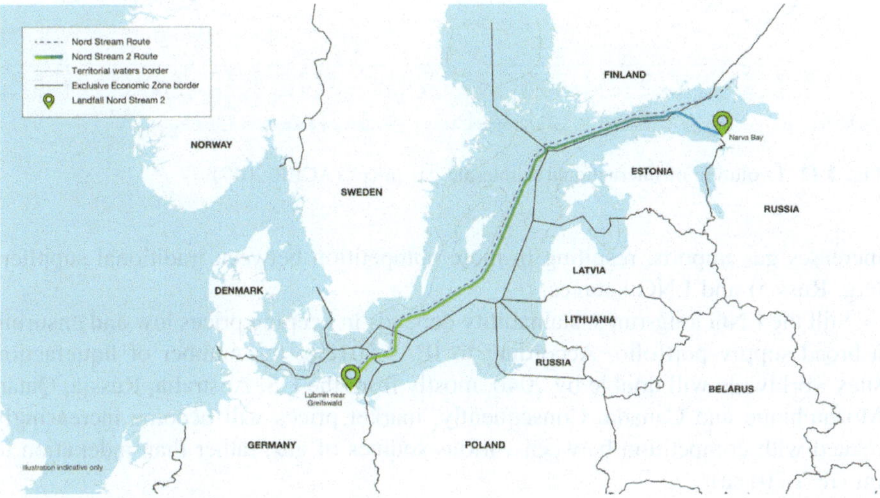

Fig. 2.40 Nord Stream 2 project (EC 2017)

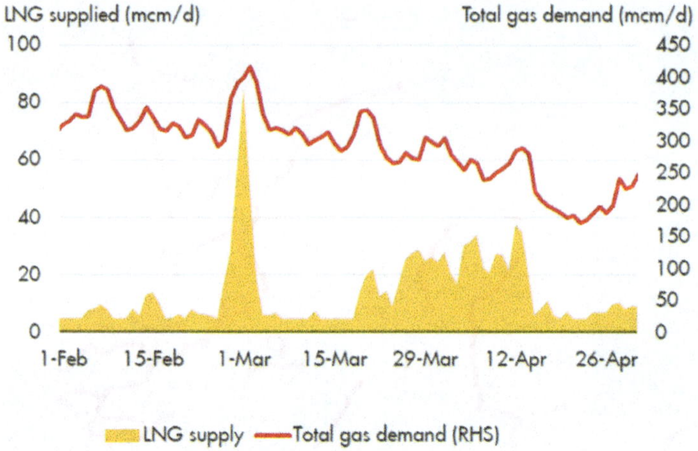

Fig. 2.41 Total gas demand and LNG supply in UK, in 2018 (Royal Dutch Shell 2019)

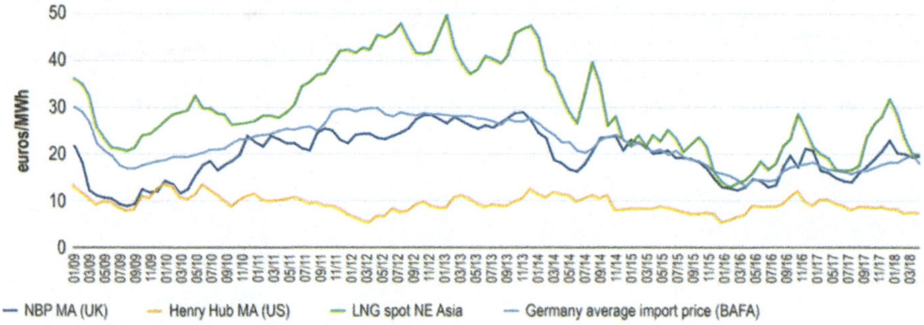

Fig. 2.42 Evolution of international wholesale gas prices (ACER 2017)

increases gas imports, resulting in more competition between traditional suppliers (e.g. Russia) and LNG sources.

Still the LNG long-run sustainability depends in keeping prices low and ensuring a broad supply portfolio. According to IEA (2018a), the number of liquefaction sites worldwide will double by 2040, mostly from the US, Australia, Russia, Qatar, Mozambique and Canada. Consequently, market prices will become increasingly related with competition between various sources of gas, rather than indexation to oil (IEA 2018a).

From the demand side of LNG, Asia is expected to keep its 70% market share in the coming decade. Yet the growth drivers are likely to shift from the traditional JKT markets (standing for Japan, South Korea and Taiwan) to the emerging economies, namely China, India, South and Southeast of Asia. Power generation idiosyncrasies (rebound of nuclear power in Japan and coal and nuclear competition

Table 2.6 Current and future LNG infrastructure in the EU and Turkey (King and Spalding 2018; ENTSOG 2014; MarketScreener 2019; EPDK 2018; Grain 2019)

Country	Terminal	Company	Send-out capacity (bcm/year)	Storage capacity (m^3)	Commissioning year	Capacity reinforcement plan
Belgium	Zeebrugge	Fluxys LNG	9.0	380,000	1987	2019 (to 560,000 m^3 and 12.0 bcm/yr)
Spain	Barcelona	Enagas	17.1	840,000	1969	N.A.
	Bilbao	BBG	8.8	450,000	2003	N.A
	Canary Islands	Gascan (developer)	1.3	150,000	2021 (expected)	N.A.
	Cartagena	Enagas	11.8	619,500	1989	N.A.
	El Musel—Gijon	Enagas	7.0	300,000	2013 (in hibernation)	2021 (to 600,000 m^3 and 8.8 bcm/yr)
	Huelva	Enagas	11.8	619,500	1988	N.A.
	Mugardos	Reganosa	3.6	300,000	2007	2023 (to 500,000 m^3 and 7.2 bcm/yr)
	Sagunto	Saggas	8.8	600,000	2006	N.A.
France	Dunkerque	Dunkerque LNG	13.0	600,000	2016	N.A
	Fos Tonkin	Elengy	3.4	155,000	1972	N.A.
	Montoir-de-Bretagne	Elengy	10.0	360,000	1980	2023 (to 550,000 m^3)
	Fos Cavaou	Fosmax LNG	8.2	330,000	2010	2020 (to 550,000 m^3 and 16.5 bcm/yr)
Greece	Revithoussa	DESFA	8.25	225,000	2000	2018
Italy	Panigaglia	GNL Italia	3.5	100,000	1971	2022 (to 240,000 m^3 and 8.0 bcm/yr)
	Porto Levante	Terminale GNL Adriatico	8.0	250,000	2009	N.A.
	OLT Offshore LNG Toscana	ECOS	3.75	135,000	2013	N.A.
	Gioia Tauro	LNG MedGas Terminal	12.0	640,000	2022 (expected)	N.A.
	Porto Empedocle	Gascan	8.0	320,000	2021 (expected)	N.A.

(continued)

Table 2.6 (continued)

Country	Terminal	Company	Send-out capacity (bcm/year)	Storage capacity (m^3)	Commissioning year	Capacity reinforcement plan
Lithuania	Klaipėda	Hoegh LNG/Klaipedos Nafta	4.0	170,000	2014	N.A.
Malta	Delimara	ElectroGas Malta	0.7	125,000	2017	N.A.
The Netherlands	Gate	Gate terminal BV	12.0	540,000	2011	N.A.
Poland	Swinoujscie	Polskie LNG	5.0	320,000	2016	2020 (to 480,000 m^3 and 7.5 bcm/yr)
Portugal	Sines	REN Atlântico	7.6	390,000	2003	N.A.
UK	Grain	National Grid	19.5	1,000,000	2005	2025 (to 1.2 million m^3 and 27.5 bcm/yr)
	Dragon	Dragon LNG	7.6	320,000	2009	N.A.
	Milford Haven—South Hook	South Hook	21.0	775,000	2009	N.A.
Turkey	Aliaga Etki	Etkiliman	5.3	145,130	2016	N.A.
	Aliaga Izmi	EgeGaz LNG	6.2	280,000	2006	N.A.
	Dortyol	Botas	5.2	263,000	2018	N.A.
	Gulf of Saros	Botas	7.3	N.I.	2019	N.A.
	Iskenderun	Botas	7.3	N.I.	2019	N.A.
	Marmara Ereglisi	Botas Petroleum Pipeline Corp.	6.2	255,000	1994	N.A.

Legend
N.A.: Not applicable
N.I.: No information

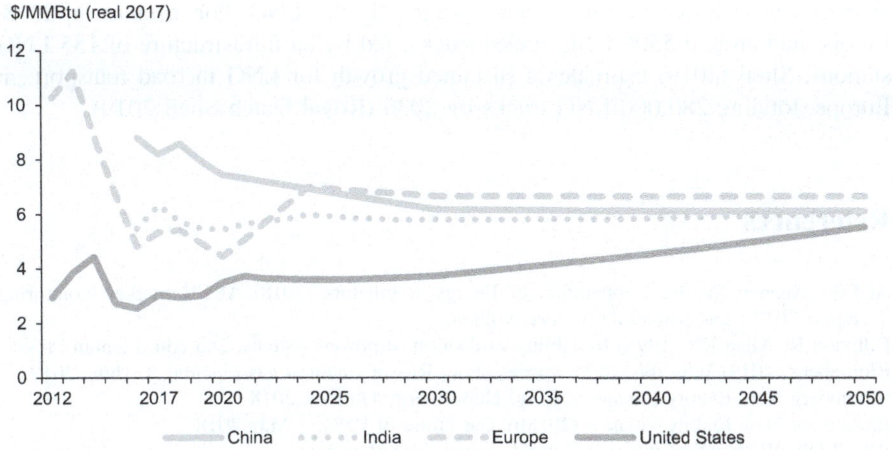

Fig. 2.43 Forecast of natural gas prices by region (Bloomberg New Energy Finance 2018a)

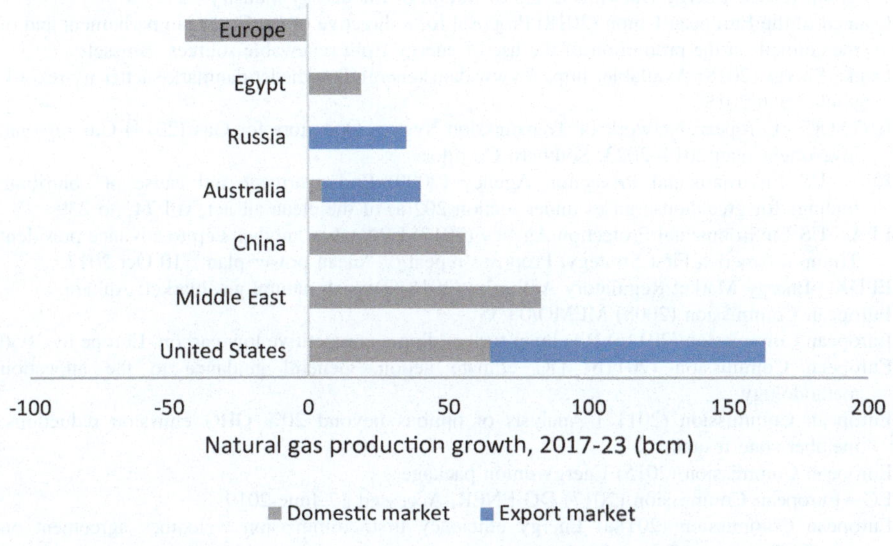

Fig. 2.44 Natural gas production growth, 2017–2023 [data from IEA (2018)]

in South Korean) will contribute to a drop in JKT's demand share of LNG from today's 49 to 25% by 2030. Alternatively, LNG demand from Asian emerging economies will likely soar to 42% by 2030 (from current 23%), mostly due to environmental policies (e.g. air quality improvement) and market reforms (Bloomberg New Energy Finance 2018b) (Fig. 2.44).

To conclude this section, one should point out that besides the power generation and gas-related uses in the industry and households, heavy-duty transport could arise as a growth avenue for gas and, specifically, for LNG. For instance, in 2018, Europe had around 5500 LNG-fueled trucks, fed by an infrastructure of 155 LNG stations. Shell (2019) estimates a sustained growth for LNG in road transport, in Europe, totaling 280,000 LNG trucks by 2030 (Royal Dutch Shell 2019).

References

ACER—Agency for the Cooperation of Energy Regulators (2018) ACER market monitoring report 2017—gas wholesale markets volume
Billinton R, Allan RN (1996) Reliability evaluation of power systems, 2nd edn. Plenum Press
Bloomberg (2019) Why the world worries about Russia's natural gas pipeline, 13 June 2019
Bloomberg New Energy Finance (2018a) New Energy Outlook 2018
Bloomberg New Energy Finance (2018b) The future of LNG, 4 May 2018
BP (2018) BP statistical review of world energy 2018, London
BP (2019) Energy outlook 2019
Carbon Tracker (2018) EU carbon prices could double by 2021 and quadruple by 2030. 26 April 2018
Citigroup (2013) Energy Darwinism: the evolution of the energy industry
Council of the European Union (2018) Proposal for a directive of the European parliament and of the council on the promotion of the use of energy from renewable sources. Brussels
Danks Energi (2018) Available: https://www.danskenergi.dk/nyheder/danmark-saetter-ny-rekord-vind. 3 Jan 2018
ENTSOG—European Network of Transmission System Operators for Gas (2014) Gas regional investment plan 2014–2023: Southern Corridor
EPA—US Environmental Protection Agency (2009) Endangerment and cause or contribute findings for greenhouse gases under section 202(a) of the clean air act, vol 74, no 239
EPA—US Environmental Protection Agency (2017) EPA takes another step to advance president Trump's America First Strategy. Proposes repeal of "clean power plan", 10 Oct 2017
EPDK—Energy Market Regulatory Authority (2018) Turkish natural gas market, Ankara
European Commission (2008) MEMO/08/35
European Commission (2011a) Roadmap for building a competitive low-carbon. Europe by 2050
European Commission (2011b) DG climate action. General guidance to the allocation methodology
European Commission (2011c) Analysis of options beyond 20% GHG emission reductions: member state results
European Commission (2015) Energy union package
EC—European Commission (2017) DG-ENER. Accessed 17 June 2019
European Commission (2018a) Energy efficiency first: commission welcomes agreement on energy efficiency, 19 June 2018
European Commission (2018b) Report from the commission to the European Parliament and the Council, EU and the Paris Climate Agreement: taking stock of progress at Katowice COP, Brussels
European Council (2014) European Council (23 and 24 October 2014). EUCO 169/14
Expresso (2019) Leilão de energia solar afunda o preço da eletricidade para níveis inéditos em Portugal, 2019 July 26
Grain LNG (2019) Available: http://grainlng.com/. Accessed 18 June 2019
GWEC—Global Wind Energy Council (2019) Available: https://gwec.net, 15 March 2019. Accessed 16 Apr 2019
Harvard Law Review (2016) The clean power plan, vol 129, no 4, p 1152

IEA—International Energy Agency (2017a) CO_2 emissions from fuel combustion: highlights. OECD, Paris
IEA—International Energy Agency (2017b) World energy outlook. OECD/IEA
IEA—International Energy Agency, Gas (2018) Analysis and forecasts to 2023
IEA—International Energy Agency (2018a) World energy outlook, Paris
IEA—International Energy Agency (2018b) World energy balances: overview, Paris
IEA—International Energy Agency (2018c) IEA Sankey Diagram. Available: https://www.iea.org/Sankey/#?c=World&s=Finalconsumption. Accessed 11 Dec 2018
IEA—International Energy Agency (2018d) Market report series: renewables 2018, Paris
IPCC—Intergovernmental Panel on Climate Change (2007) Fourth assessment report: climate change
IPCC—Intergovernmental Panel on Climate Change (2018) Global Warming of 1.5 °C
IRENA—International Renewable Energy Agency (2019) Renewable capacity 2019
Jinping X (2015) Work together to build a win-win, equitable and balanced governance mechanism on climate change, Paris
King & Spalding (2018) LNG in Europe 2018: an overview of LNG import terminals in Europe
Lannoye E, Flynn D, O'Malley M (2012a) Evaluation of power system flexibility. IEEE Trans Power Syst 27(2):922–931
Lannoye E, Flynn D, O'Malley M (2012b) Assessment of power system flexibility: a high-level approach. In: IEEE power and energy society general meeting
MarketScreener (2019) EU-U.S. Joint Statement: Liquefied Natural Gas (LNG) imports from the U.S. continue to rise, up by 181%, 8 March 2019
Moreira da Silva M, Pastor R, Shi T, Zhao L, Ye J (2015) Siting and sizing dispersed energy storage in power transmission networks. In: IEEE green energy and systems conference, Long Beach, CA
NERC—North American Electric Reliability Corporation (2010) Special report: potential reliability impacts of emerging flexible resources
OECD (2012) Green growth and the future of aviation. In: Paper prepared for the 27th round table on sustainable development
OECD (2017) Investing in climate, investing in growth
Official Journal of the European Union (2009) Directive 2009/29/EC of the European Parliament and of the Council
REN—Redes Energéticas Nacionais (2013) Information Centre—Electricity, 20 April 2013. Available: http://www.centrodeinformacao.ren.pt/PT/InformacaoExploracao/Paginas/EstatisticaDiariaDiagrama.aspx. Accessed 13 May 2019
REN—Redes Energéticas Nacionais (2018) Available: www.ren.pt. Accessed 3 May 2018
Reuters (2017) US submits formal notice of withdrawal from Paris climate pact. World News, 4 Aug 2017
REVE (2019) Available: https://www.evwind.es/2019/01/25/the-wind-power-beats-record-of-daily-production-with-43-2-of-the-total-in-spain/65909., 25 Jan 2019
Royal Dutch Shell (2019) Shell LNG Outlook 2019
RTE—Réseau de Transport d'Electricité (2016) Carbon price signal—impact analysis on the European electricity system, Paris
Song S (2018) Here's how China is going green. World Economic Forum, 26 April 2018
Stern N (2006) Stern review: the economics of climate change
Stiglitz J (2006) Making globalization work, W.W. Norton & Company, Inc
The New York Times (2017) Q.&A.: The Paris Climate Accord, 31 May 2017
The White House (2013) The president's climate action plan, Washington
The World Bank (2018) The World Bank. Available: https://www.worldbank.org/. Accessed 16 Nov 2018
UNFCCC—United Nations Framework Convention on Climate Change (2008). Available: www.unfccc.org. Accessed 29 Oct 2008
UNFCCC—United Nations Framework Convention on Climate Change (2018a). Available: https://unfccc.int. Accessed 16 Nov 2018

UNFCCC—United Nations Framework Convention on Climate Change (2018b) China meets 2020 carbon target three years ahead of schedule. United Nations, 28 Mar 2018
United Nations Development Programme—UNDP (2019) Human development data. Available: http://hdr.undp.org/en/data. Accessed 17 Apr 2019
World Nuclear Association (2011) Comparison of lifecycle greenhouse gas emissions of various electricity generation sources, London

Chapter 3
Regulation and Business Models

3.1 Fundamentals of Regulation Theory

Regulation is a concept often misunderstood, misused or even perceived as a daunting action of a central entity—headquartered in Washington, Brussels or Beijing—over the industry. Economic regulation, though, seeks to maximize the general welfare, including consumers and investors in the same objective function. Regulation benefits the consumers, by preventing unreasonable prices drawn by monopolies and oligopolies. On the other hand, regulation protects the investors from the State, by ensuring cost recovery (i.e. prices reflecting costs). This is true not only for the energy sector, but also for a myriad of industries such as water, telecommunications and railways.

Energy utilities are entailed in a unique political, economic and social context. Power and gas companies are rooted on physical assets, mostly understood as a natural monopoly. In view of this, every investment plan, shareholder change and corporate reorganization process is often preceded by political discussion. Additionally, since power and gas utilities are capital intensive businesses with direct link to the consumer/citizen—who pays for the services—energy planning (e.g. capacity expansion) and system operation (e.g. quality of service) are highly scrutinized. Consequently, the utilities' business is involved in a permanent tension between the company (i.e. investor) and the consumer's interests (one could also add other stakeholders, such as environmental organizations).

At this initial stage it is worth providing a simple description of a natural monopoly. Decker (2014) suggested that "in certain conditions, it is most cost efficient if a single firm, rather than two or more firms, produces a specific set of outputs". Typically, natural monopolies encompass the following attributes: (i) economies of scale; (ii) capital-intensity; (iii) non-storability with fluctuating demand; (iv) locational specificity generating location rents; (v) essential for the community; and (vi) directly connected to customers (Farrer 1902). If one adds the concept of competition—which promotes costs minimization—natural monopolies

M. Moreira da Silva, *Power and Gas Asset Management*, Lecture Notes in Energy 72, https://doi.org/10.1007/978-3-030-36200-3_3

inevitably lead to market failure. In the absence of competition, natural monopolies face no pressure for service optimization (let alone, cost-cutting) and could result on excessive costs.

The idiosyncrasy of natural monopolies—and the market failure risk—sets up the case for designing a regulatory framework with checks-and-balances amongst different players: the government (concession grantor); the utility (concessionaire); the end-users (e.g. consumers); and the regulator. Although there is a quasi-consensus for this regulatory architecture, the regulator's degree of independence and power varies from country to country. This is mostly related with the public or private ownership of the utilities. Usually, countries where grid companies are private-owned have an independent and empowered regulatory authority. Therefore, privatization processes of energy utilities are coupled with regulatory reforms, in order to prevent windfall profits from the new investor (who, naturally, seeks to maximize the return on investment) and ensure security of supply, quality of service and competition (in the supply and retail activities). On the other hand, countries where energy utilities are state-owned have a rather unclear distinction between the roles of the government, the utility and the regulator (who lacks independence, acting as another government body). In this case, the state is both concession grantor, shareholder of the concessionaire and regulator.

Besides the utilities ownership discussion (which is out of the scope of this book), regulation models depend of the sector framework, whether there is a vertical integrated company, ownership unbundling or liberalized access to the grid. Regardless the value chain arrangement, competition is deemed as rather efficient than regulation. In view of that, regulation should be confined to the grid (the natural monopoly), whereas competition must prevail in the supply and retail activities. As stated by Newbery (1999), "liberalization, if successful, puts competitive pressure on the incumbent utility and, if unsuccessful, creates pressure for regulatory reform". Joscow (2008) and Batle (2013) have also proposed guidelines towards liberalized and competitive power markets—partly applicable to the gas sector—as follows.

1. Set up of an independent regulatory authority.
2. Privatization (to increase performance and prevent political agendas).
3. Unbundling (separation of a vertical integrated company): spin-off of competitive activities (supply and retail) from the natural monopolies (transmission and distribution grids) that must be regulated.
4. Horizontal restructuring of the generation business, to create an adequate number of competing generators to mitigate market power.
5. Set up of an Independent System Operator (ISO), responsible for: system adequacy and security; energy dispatching; grid planning and operation; and open access to the wholesale market and the transmission grid.
6. Design and establishment of a wholesale market, to put in place competition in the supply side.
7. Enable access to the distribution grid, in order to accelerate competition in the retail side.

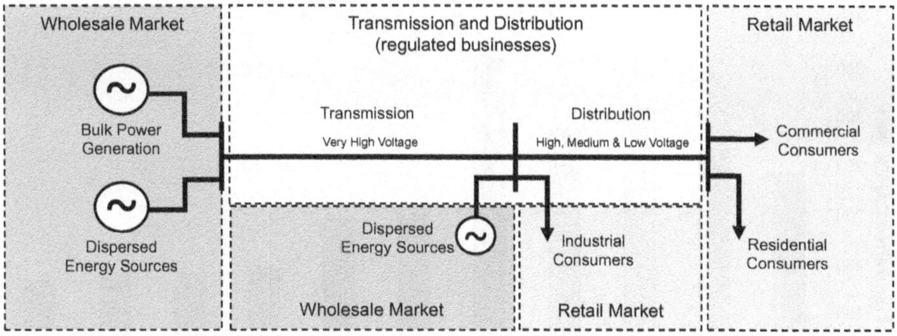

Fig. 3.1 Value chain of the power sector

8. Develop the market for active demand-side response.

Taking this recipe, liberalization of power and gas utilities has spread around the world, initially in the UK, Norway, Chile, Argentina, some states of the US (e.g. Texas) and rather recently in the EU. The energy liberalization in the EU was inspired by the UK case, though addressing not only competition but also security of supply and sustainability. The EU opted for the ownership unbundling of power and gas utilities, in order to ease competition in the "wireless" businesses (supply and retail). The value chains of power and gas services are presented in Figs. 3.1 and 3.2, following the incumbent vertical separation.

One example of the virtues of privatization is found at Portugal's TSO, REN—Redes Energéticas Nacionais. REN was set up as an independent company, in the year 2000, following the vertical separation of the incumbent (EDP). Yet only in 2012 the company has become privately controlled, after the second reprivatization process. Since REN's regulatory model is based on a rate of return of the invested capital, the company is incentivized to transfer CAPEX to the regulated asset base

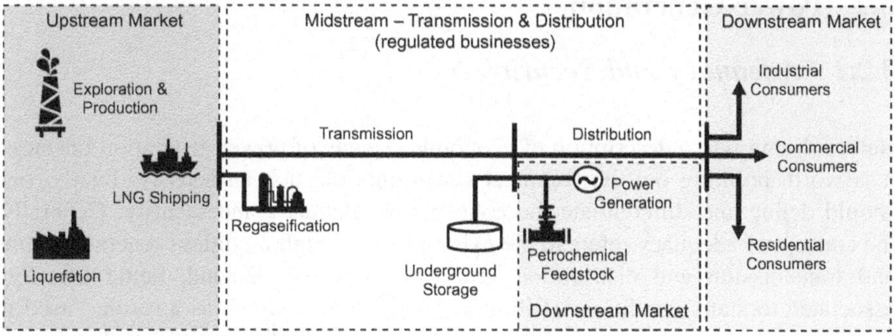

Fig. 3.2 Value chain of the natural gas sector

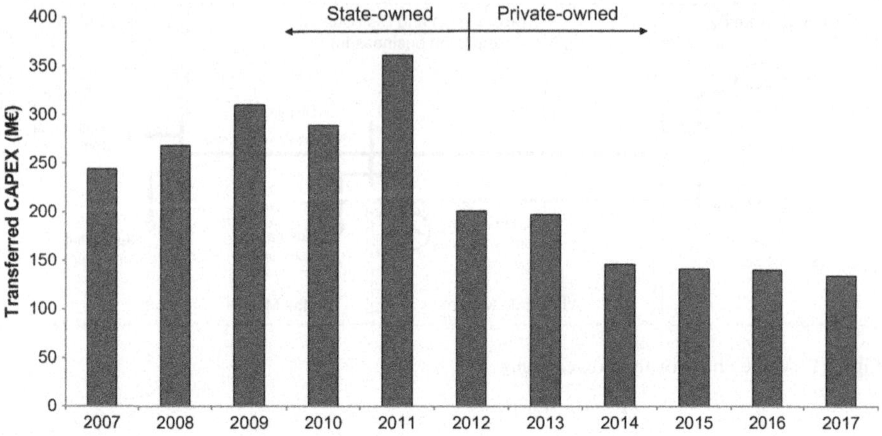

Fig. 3.3 Portugal's TSO grid investment (REN 2008, 2009, 2010, 2011, 2012, 2013a, b, 2014, 2015, 2017, 2018)

(RAB). In truth, after REN's privatization, the investment has slumped due to a different positioning of the government and the regulator (with augmented power), leading to higher standards for accepting the company's investment projects (through cost-benefit analyses). From 2007–2011 the average transferred CAPEX to operation was 294 million €/year, whereas from 2013–2017 this figure was halved to 152 million €/year (Fig. 3.3). Although the financial discipline after the entry of private shareholders, the power quality and technical performance were kept high (average interruption time bellow 0.4 min, half of the substation outages per circuit-end and 2/3 of line outages per 1000 circuit km, comparing to the peers' average), placing REN as "best in class" among international TSOs. In this case, privatization brought transparency and cost-effectiveness of the power transmission activity, without jeopardizing security of supply and reliability of the grid.

3.2 Power Generation

3.2.1 Adequacy and Security

Before diving into a description of economic aspects of power generation business, it is worth pointing out the technical constraints of such an activity. Firstly, one should define and differentiate the concepts of adequacy and security. Generally, the concept of adequacy refers to the existence of enough facilities (generation units and transmission and distribution grids) to meet the demand, being therefore associated to static conditions (Billinton 1966). Adequacy is, as a result, "used to describe a system state in which the actual entry to and departure from that state is ignored and is thus defined as a steady-state condition", as stated by Billinton and Allan (1996).

On the other hand, security is concerned to the ability of the system to respond to disturbances arising within the system. These disturbances include conditions causing local and broad effects, as well as the loss of major generation and transmission lines (Billinton 1966). Typically (but not exclusively), security is related to the dynamic process that occurs when the system transits between one state and another state (Billinton and Allan 1996). In this final state, both equality and inequality constraints are verified.

Adequacy and security are, however, correlated. That is to say, a system with large amount of reserve capacity has a higher flexibility to react to unforeseen disturbances. On the other hand, when facing a power shortage, a given system with limited planning reserves can still be operated in a secure way, whereas a system with a larger reserve can be operated insecurely (Martins et al. 2012).

(i) Hierarchical levels of the system

The most common approach for analyzing a power system is through the identification of its main functional zones, namely: generation systems; transmission systems; and distribution systems.

Billinton (1966) presented the concept of hierarchical levels (HL), aiming to identify group functional zones of power systems. The referred author set out three levels:

- HL1: consists of generation facilities and their ability to satisfy the demand;
- HL2: refers to the composite generation and transmission system, and its ability to deliver electricity to supply points;
- HL3: refers to the complete system, including the distribution grid and its ability to satisfy the capacity and energy demands of consumers (these studies are rather complex, owing to the scale of the problem).

(ii) Reliability Assessment of Power Generation

The problem of defining the necessary amount of generating capacity is divided in two areas of study: static capacity requirements; and operating capacity requirements. The static capacity area performs a long-run assessment of the system requirements, being considered as the installed capacity that must be planned and built in advance. The static reserve must provide to the system enough capacity to handle outages, scheduled maintenance and load growth.

The basic approach to evaluate the adequacy of a certain generation configuration consists of three parts: generation model; load model; and risk model (Fig. 3.4). The generation and load models are combined (convolved) to form the appropriate risk model (Billinton and Allan 1996).

When it comes to the operating capacity study, it is concerned with short-term assessment of the actual required capacity to satisfy the demand. System operators have developed techniques for managing the variability and uncertainty of demand and conventional generation, by establishing operating reserves. The term

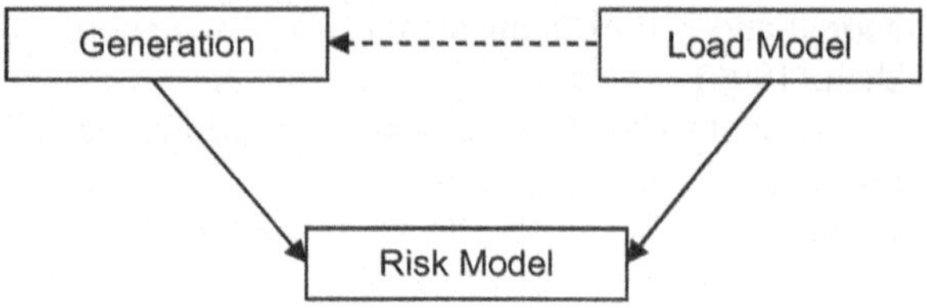

Fig. 3.4 Generating capacity reliability evaluation

"operating reserve" was defined in Holttinen et al. (2012) as the "active power capacity that can be deployed to assist with generation and load balance and frequency control". The major difference between static and operating capacity evaluation stands for the period of time under analysis (Billinton and Allan 1996).

The integration of renewable energy sources has led to several impacts on the power systems operation, namely on security of supply, such as the required mobilization at peak load periods (due to lack of RES) and the response to sudden generation drops (owing to rapid changes in the resource). This and other drivers increase the need for operating reserve. Although deterministic approaches have very attractive characteristics (e.g. simple implementation, easy understanding, assessment and judgment), the perception of many planning engineers that past experience, in addition to some known critical situations, is enough to assess system risk conditions is not valid. Furthermore, track-record with renewable sources, such as wind power, is scarce. Yet the principles of some deterministic standards (e.g. "N-1" criterion) are useful. Methodologies based on probability concepts are useful to assess the power systems performance, having been applied to generation and transmission capacity planning, operating reserve assessment, distribution systems, etc. Conventional probabilistic indices (described subsequently) are already widely used, and Well-Being Analysis has been built combining the deterministic perception with probability concepts (Matos et al. 2009).

Chronological or sequential Monte Carlo simulation has been used for generating system well-being analysis. Matos et al. (2009) presented an application of chronological Monte Carlo simulation to evaluate the reserve requirements of generating systems, considering renewable energy sources. The referred work studied the behavior of reliability indices (conventional and well-being), when a major portion of the energy sources is renewable (i.e. mainly hydro, wind and mini-hydro power sources). Sequential simulation identifies all chronological aspects, being capable to represent equipment aging process, time varying loads, and spatial and time correlations. According to the referred authors, chronological Monte Carlo simulation is very suitable owing to its flexibility, since it allows representing non-exponential residence times, meaningful when dealing with chronological processes. Additionally, chronological Monte Carlo simulation is an effective mean to adequately model natural uncertainties of RES (i.e. hydrologic inflow sequences, wind speed variations, etc.).

(iii) Reliability Indices

In static capacity evaluation problems, the basic generating unit parameter to be used, is the probability of finding the unit in forced outage. This probability is known as Forced Outage Rate (FOR) and expressed as follows (Billinton and Allan 1996).

$$Unavailability = FOR = U = \frac{\lambda}{\lambda+\mu} = \frac{r}{m+r} = \frac{r}{T} = \frac{f}{\mu} = \frac{\sum[down\,time]}{\sum[down\,time] + \sum[up\,time]} \quad (3.1)$$

$$Availability = A = \frac{\mu}{\lambda+\mu} = \frac{m}{m+r} = \frac{m}{T} = \frac{f}{\lambda} = \frac{\sum[up\,time]}{\sum[down\,time] + \sum[up\,time]} \quad (3.2)$$

where,

λ is the expected failure rate;
μ is the expected repair rate;
m is the mean time to failure, which is equal to $1/\lambda$;
r is the mean time to repair, which is equal to $1/\mu$;
$m + r$ is the mean time between failures, which is equal to T;
f is the cycle frequency;
T is the cycle time, which is equal to $1/f$.

Figure 3.5 presents the concepts of unavailability and availability for the two-state model.

The capacity outage probability table (COPT) is an array of capacity levels and the corresponding probabilities. The units can be combined applying probability concepts and this methodology can be extended to build a useful recursive technique (Billinton 1970).

As alternative to the classical recursive technique, one can find other methodologies for the COPT computation. In cases where the system is very large, the discrete distribution of system capacity outages can be approximated by a

Fig. 3.5 Two-state model (Billinton and Allan 1996)

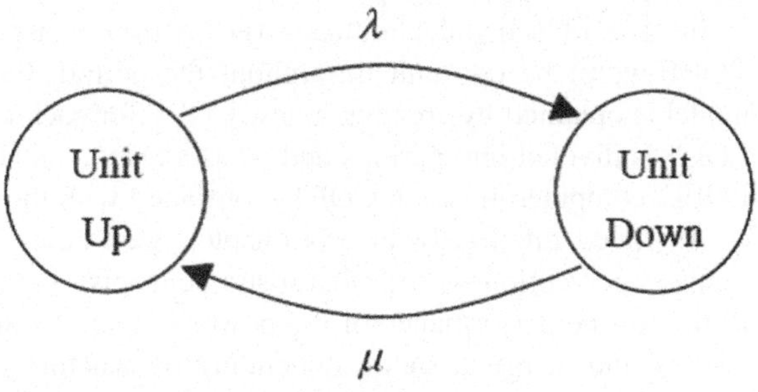

continuous distribution (Bhavaraju 1974), which approaches the normal distribution as the system size increases. However, the results achieved with this methodology are less accurate comparing to the obtained using the recursive technique (Allan and Takieddine 1977). In addition, Schenk and Rau (1979) proposed a Fourier transform method in order to improve the accuracy of the continuous model, though it is only compared with the recursive technique when the system is sufficiently large (Schenk and Rau 1979). This approach is especially inaccurate for systems with hydro units with low FOR (Stremel 1981).

Another alternative approach consists of transforming the unit capacity tables into the frequency domain through fast Fourier transforms and to convolve using a point by point multiplication. An inverse fast Fourier transform can be used to generate the final COPT (Allan et al. 1981). This method is considerably faster than the direct recursive technique and leads to more accurate results than the Fourier transform method (Billinton and Allan 1996).

The most common indices used for the assessment of generation adequacy are the Loss of Load Expectation (LOLE), the Loss of Energy Expectation (LOEE) and the Loss of Load Frequency (LOLF). The LOLE is the expected number of days (hours) in a specified period in which the daily peak load (hourly load) exceeds the available generating capacity. The LOEE is the expected unsupplied energy due to generating inadequacy and incorporates the severity of the deficiencies. The LOLF is the expected frequency of encountering a generation deficiency in a given period (Billinton and Huang 2005; Billinton and Chu 1992). Most electric power utilities adopt the LOLE index in their generation system planning.

For LOLE calculation, the generation model is convolved with the load model to produce the risk index. There is a wide range of load models, but the simplest (and widely used) is the daily peak load, in which the individual daily peak loads are arranged in descending order, to form the daily peak load variation curve. Alternatively, the model is named load duration curve, when the individual hourly loads are used.

The loss of load index is then obtained by combining the COPT with load characteristics. It should be emphasized that a loss of load occurs when the generating capacity remaining in service is exceeded by the load level. For the case of LOLE index, the individual daily peak loads are combined with the COPT, to get the expected number of days in which the daily peak load will exceed the available capacity, in the specified period.

In case of scheduled outages (i.e. periodic inspection and maintenance), the COPT won't be constant throughout the period, therefore the modified capacity model is obtained by creating a new COPT for each capacity condition. The annual LOLE is divided into periods and is, as a result, calculated as the sum of individual LOLE, computed for each COPT (combined with the period load model) (Fig. 3.6).

As stated previously in this chapter, well-being analysis combines the deterministic perception with probabilistic concepts, and provides additional characterization of the performance of the power system, by splitting the success states into healthy and marginal states, depending on whether or not a deterministic rule for reserve is satisfied. The specified value for secondary reserve or the largest available

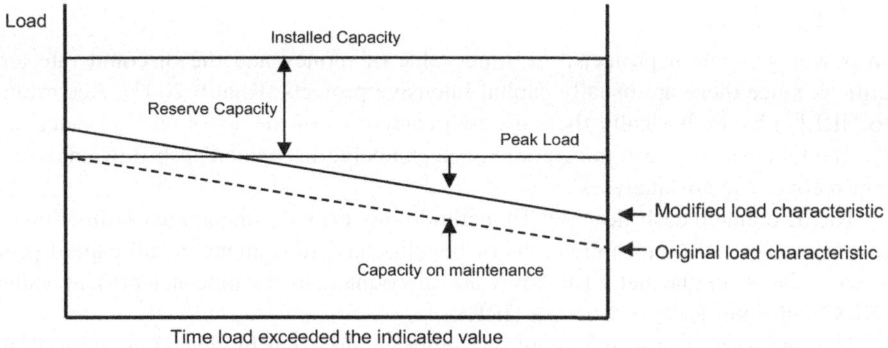

Fig. 3.6 Approximate method for including maintenance [adapted from (Billinton and Allan 1996)]

unit in the system are typical thresholds used for this purpose. For instance, in the first case, the state is considered healthy if the margin between available generation and load is greater than the required secondary reserve (Lopes et al. 2008).

Well-being analysis encompasses the next indices (Billinton and Fotuhi-Firuzabad 1994; Billinton and Karki 1999; Leite da Silva et al. 2004; Leite da Silva et al. 2007).

- EH—expected healthy hours, which is the expected number of hours in a period (e.g. year) the system will stay in healthy states.
- EM—expected marginal hours, which is the expected number of hours in a period (e.g. year) the system will stay in marginal states.
- FH and FM—expected frequency associated with healthy and marginal states, respectively.
- DH and DM—expected duration of system residing in healthy and marginal states, respectively.

3.2.2 Economics

The life cycle cost of electricity generation is usually defined as the discounted lifetime cost of ownership of using a generation asset, converted into an equivalent unit cost of generation (Mott MacDonald 2010). There are two main components of life cycle costs, as follows.

- Investment (or capital) costs of bringing the asset to a point of operation. These are usually called Capital Expenditures (CAPEX).
- Operation Costs, which include the fixed costs (of keeping the plant available to generate) and the variable of costs of operation. These are usually called Operation Expenditures (OPEX).

(i) Capital Costs

In power generation projects, the time value of money and the discount rate are critical, since these are usually capital-intensive projects (Khatib 2003). According to NREL (2011), basically there are two categories of methodologies to calculate the capital costs of power generating assets, namely: discounted cash-flow analyses; and recovery factor analyses.

The discounted cash-flow (DCF) methodology provides the annual estimation of revenues, expenses, tax obligations or benefits, and repayments to all capital providers. These annual net cash flows are discounted to a single net present value (NPV) and internal rate of return (IRR).

The recovery factor methodology replaces the year-by-year free cash flow forecast of the discounted cash-flow analysis, by converting capital costs into an annual figure. Therefore, the annual capital cost is calculated by multiplying the initial investment by the recovery factor. A set of alternative methods fall into the recovery factor analysis category, such as: capital recovery factor (CRF); fixed-charge rate (FCR); and Economic Carrying-Charge Rate (ECCR).

Table 3.1 outlines the described methodologies for the capital cost calculation, including also other investment analysis tools (simple payback and profitability index method).

The capital costs of power plants entail the next set of components:

- The main plant and equipment package, often referred as engineering, procurement and construction (EPC) price.
- Infrastructure/connection costs, including power, fuel and cooling system (these could also be included in the EPC price).
- Development costs, including permitting, advisory services and land rights.
- Interest and funding cost during construction.

It is possible to compare the power generation technologies, according to their capital costs. Mott MacDonald (2010) presented the next power technologies hierarchy in terms of capital costs: nuclear is more expensive than coal; coal is more expensive than oil fired plant; fired boiler-steam plant is more expensive that combined cycle gas turbine (CCGT); and CCGT costs circa 50% more than an equivalent open cycle gas turbine. Regarding renewable energy sources, biomass combustion-based plant could be deemed as an expensive version of a coal plant. For wind power, offshore projects have higher capital costs, owing to more complex foundations, assembly and electrical cable connection to shore.

The capital costs differences among generation technologies are led by a combination of the complexity of the technology and its energy density. Energy technologies that generate more energy per square meter, usually need less material inputs than those with lower energy densities. Furthermore, technologies with complex designs and production processes, often have higher capital costs per installed power, than less sophisticated technologies. In addition, there are other characteristics that influence the capital cost, such as the fuel and residue handling requirements, and control and safety systems (Mott MacDonald 2010).

Table 3.1 Methods for capital cost calculation (NREL 2011)

Methodology	Description	Calculation
Discounted cash-flow	Discounts to present value the estimated annual cash flows to equity investors; provides either before-tax or after-tax results	Initial equity investment plus net present value of free cash flow to equity over the project life; internal rate of return of investment and cash flows for a specified period
Capital recovery factor	Amortizes an investment into a stream of equal annual payments; provides pre-tax results; also called "annuity method"	Sums weighted average cost of capital and depreciation annuity
Fixed-charge rate	Calculates the portion of a project's year-one revenue requirement, attributable to fixed costs; provides after-tax results	Sums annual weighted average cost of capital, tax, depreciation, and fixed overhead
Economic carrying-charge rate	Amortizes all fixed costs to produce a stream of annual payments that increase at a constant rate; provides after-tax results	Sums year-one weighted average cost of capital, tax, depreciation, and fixed overhead to derive year-one cost of energy
Simple payback	Estimates the number of years necessary to recover an initial equity investment; provides before-tax results	Initial equity investment/ annualized cash flow to equity
Profitability index method	Indicates the efficiency of invested capital; used to rank projects based on net present value per money invested	Net present value/total installed cost

The main cost drivers for mature power technologies are market conditions and commodity prices. These technologies are usually called nth of a kind (NOAK), while the units still at a learning stage are called first of a kind (FOAK). Consequently, FOAK encloses additional risks than NOAK, related to the implementation of new technologies, new construction techniques and supply chain management. Traditionally, these risks are carried by the original equipment manufacturer (OEM), the EPC contractor and the developer.

(ii) Operation Costs

Within the operation costs, one should take into consideration a set of fixed costs, namely: workforce allocation; preventive and corrective maintenance actions; inspections to the facilities; asset replacement and refurbishment; property taxes; and insurance.

Concerning the variable costs of operation, these include fuel consumption and carbon emissions. Fuel costs are determined by the type of fuel, heat rate and fuel prices. Power plants can be categorized either as being expensive machines that convert low-cost energy into electricity (e.g. renewable energy sources), or low-cost engines that convert expensive fuels into electricity (i.e. fossil fuels). For the expensive fuel converters, the major cost drivers are the efficiency of fuel conversion and the fuel price (Mott MacDonald 2010).

Concerning renewable energy sources, the life cycle cost of the projects is dominated by capital costs, rather than operation expenditures. However, the annual fixed and variable costs also impact a project's economics. According to NREL (2011), the annual OPEX of a renewable energy unit includes: project management; insurance; property taxes; permit maintenance; site maintenance; land lease or royalty payments; and others. Since many renewable energy sources are still FOAK technologies, generally these are given incentives that take many forms: tax credits and accelerated depreciation deductions; production incentives; rebates; and grants (both taxable and non-taxable). Traditionally, renewable energy facilities receive specific feed-in-tariffs (FIT). It is possible to find out wide FIT mechanisms across the world, such as the generation-cost models adopted in Germany, the Netherlands, Ontario (Canada), Vermont (US), Florida (US) (NREL 2011), and also Spain and Portugal.

Regarding non-renewable technologies, the life cycle cost of energy is dominated by fixed and variable costs. Within variable costs, fuel costs are linked to fuel consumption (which depends of the plant efficiency and power output in each period of time) and to fuel price.

Additionally, one should also consider the shut-down and start-up costs, which have an economic impact due to: fuel waste; extra maintenance; and additional feed water and energy for heating. The start-up and shut-down costs are also unpleasant from a social, technical and environmental perspective, since they are stressful to operators, reduce the effective life of generating units (owing to heating and cooling cycle, pressurization and decompression of boilers, etc.) and the GHG emissions increase during the transient period of start-up and shut-down (Viana 2004). Usually, shut-down costs are significantly lower than start-up costs.

Regarding carbon emissions, the generation companies in the EU ought to comply with the Emissions Trading System, as explained in Sect. 2.2.2. According to the European Commission, the allocation process for the emission allowances starts with the Community-wide and fully harmonized Implementing Measures (CIMs) (European Commission—DG Climate Action 2011). Based on the CIMs, the Competent Authorities calculate the preliminary annual allocation on a sub-installation level and each Member State submits to the EC the list of all incumbent installations—covered by the ETS Directive within its territory—and any free allocation to each installation. This list is called National Implementation Measures (NIMs).

The auction format is based on a single-round, sealed bid, uniform price auction (European Commission 2010). During a single bidding window of the auction, bidders can place any number of bids, each specifying the number of allowances they would like to buy at a given price. Directly following the closure of the bidding window, the auction platform will determine and publish the clearing price, at which demand for allowances equals the number of allowances offered for sale in the concerned auction (Fig. 3.7). Successful bidders are the ones who place bids for

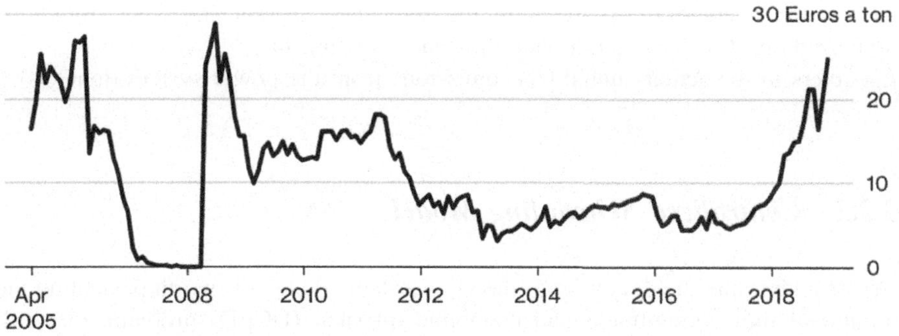

Fig. 3.7 Price of EU allowances (ICE 2019; Carr 2018)

allowances at or above the clearing price. All successful bidders pay the same price, regardless the price they specified in their bids.

To limit the impact of auctions on the secondary market, the auctions were designed to be relatively frequent. Therefore, auctioning works if the clearing price mirrors the price in the secondary market (Zapfel 2011).

The secondary markets operate carbon allowances derived from the primary market, such as European Union Allowances (EUA), Certified Emission Reductions (CER) and Emission Reduction Units (ERU), traded in major markets like ECX, BlueNext, EEX and GreenX. The secondary markets can trade allowances as spot, futures, organized and over-the-counter (OTC). Therefore, the generation companies should define adequate hedging strategies, such as a "single-shot" bid (acquiring the total amount of required allowances), sequential transactions according to needs (weekly or monthly) or a diversified strategy by combining auctions and futures (to purchase not only short-run allowances but also mid-run products) (Matos 2011).

Taking into consideration this overview on the EU ETS, at the end of each trading period installations must surrender allowances equivalent to their emissions. Thus, if a GENCO (electricity Generation Company) keeps its emissions below the emissions cap, it can sell the excess allowances. But if the GENCO has emitted more CO_2 than its cap, it will have to buy further allowances in the market.

The carbon price should be combined with the difference between the annual CO_2 emissions from the power system and the corresponding cap, as described underneath (Moreira da Silva 2013).

$$TAC_{CO_2} = \theta \times CP_1 + (E_{PS} - \theta) \times CP_2 \tag{3.3}$$

where,

TAC_{CO_2} stands for the annual carbon-based costs (€);
θ is the annual allocation of CO_2 allowances (cap), received from the auctioning (tonCO_2);
CP_1 is the average auctioning price of allowances in the EU ETS (€/tonCO_2);

CP_2 is the average EUAs price in the secondary market (i.e. spot, forwards, organized and OTC) for additional allowances (€/tonCO_2);
E_{PS} refers to the actual annual CO_2 emissions from the power system (tonCO_2).

3.2.3 Centralized Scheduling Model

Before addressing the dynamics of the wholesale markets, it is worth presenting the traditional unit commitment and economic dispatch (UCED) problem, that still rules the generation scheduling where the state owns and operates the vertical integrated utility.

Prior to tackling the economic dispatch problem, the unit commitment problem should be solved, because it is necessary to know in advance which are the units that will be operating in the interval under analysis, in order to proceed to their output, set up by the economic dispatch model. The unit commitment is a crucial sub-problem for scheduling generation units to be "on" or "off", during each interval of the scheduling period. This problem has a set of constraints, such as: meeting the demand; ensuring minimum value of spinning reserve; and complying with minimum and maximum limits of the unit.

The UCED problem has been studied for long time, moving from simple calculations to rather sophisticated methods (Li et al. 1997). There is a set of available methods for UCED, which can be grouped as follows: extensive enumeration; priority list (merit order scheduling); dynamic programming and its variants; Lagrangian relaxation; branch-and-bound method; linear programming; expert systems/artificial neural networks; simulated annealing; and genetic algorithms (Sen and Kothari 1998).

The operation costs of non-renewable energy sources are determined by the Unit Commitment and Economic Dispatch, for the corresponding generation diagram of the time under analysis.

The Unit Commitment and Economic Dispatch problem is applied to a typical daily load profile of a given power system. The UCED is divided into 24 sub-problems (one sub-problem per period) and for each period of time:

- the units are assigned to be "on" or "off";
- the power output of each unit that is "on" is calculated.

Taking into account the relevance of start-up and shut-down costs, the UCED is de facto a multi-period problem, being impossible to solve it in a separate way for each period.

The basic thermal UCED problem is modelled as a single objective problem that aims at minimizing total operation costs. Fuel costs are linked to the fuel consumption of each unit, which depends of the power output in each period and type of unit (fuel oil, diesel, etc.). The functions that represent the fuel costs are, generally, non-continuous and non-convex. Nevertheless, since non-convexity of such functions prevents the use of conventional optimization techniques, polynomial interpolation is often applied and a linearization can be made to describe the fuel consumption at different power levels, for each unit type, as referred in the literature (Matos 2007; Mendonça et al. 2004; Stoft 2002).

When it comes to start-up costs and shut-down costs, these are incurred each time a power plant is turned "on" and "off", respectively. The start-up costs depend on the latest period the unit was operating and, for steam units, on whether the machine was kept hot, or not, while the "off" period. Concerning shut-down costs, these can be represented as a constant, being lower than start-up costs, as referred in Viana (2004).

The classical UCED problem has, then, the following formulation and corresponds to the minimization of power generation operation costs in a 24 h period (Wood and Wollenberg 1996; Huang et al. 1997).

$$Minimize \sum_{i=1}^{h}\sum_{k=1}^{n}\left[u_{i,k}\times\left(F\left(P_{i,k}\right)+C_k^u\times\left(1-u_{i-1,k}\right)\right)+\left(1-u_{i,k}\right)\times u_{i-1,k}\times C_k^d\right] \tag{3.4}$$

Subject to:

$$\sum_{k=1}^{n} u_{i,k}\times P_{i,k}=P_i^L,\ \forall i \tag{3.5}$$

$$\sum_{k=1}^{n} u_{i,k}\times P_k^{max}\geq P_i^L+SR_i,\ \forall i \tag{3.6}$$

$$u_{i,k}\times P_k^{min}\leq P_{i,k}\leq u_{i,k}\times P_k^{max},\ \forall i,k \tag{3.7}$$

$$u_{i,k}=1 for \sum_{i=i_{su}}^{i-1} u_{i,k}<MUT_k,\ \forall k \tag{3.8}$$

$$u_{i,k}=0 for \sum_{i=i_{du}}^{i-1}\left(1-u_{i,k}\right)<MDT_k,\ \forall k \tag{3.9}$$

where,

$u_{i,k}$ is the status (1—on; 0—off) of unit k, at period i;
F refers to the fuel and fixed costs in period i, of unit k with $P_{i,k}$ of power output;
$P_{i,k}$ is the power output from unit k at period i;
h is the number of periods;
n is the number of units (non-renewable generators);
C_k^u refers to the costs of starting-up unit k;
C_k^d represents the costs of shutting-down unit k;

P_k^{max} is the maximum power output of the fossil-fired unit k;
P_k^{min} is the minimum power output of the fossil-fired unit k;
SR_i is the spinning reserve at period i;
MUT_k is the minimum up-time of unit k;
MDT_k is the minimum down-time of unit k;
i_{su} is the hour at which the unit k is started up;
i_{sd} is the hour at which the unit k is shut down.

The described formulation is, however, inadequate for power systems with renewable generation and electricity storage facilities, including EVs' V2G operation. For the UCED with renewable generation, such as wind, the renewable output is typically subtracted from demand, producing the "net load," which can then be used to compute the thermal generator requirements (Palmintier and Webster 2011). The modified formulation for UCED is now presented, by integrating the power output from renewable energy sources, centralized storage and V2G (if exist), for a typical day of operation (Moreira da Silva 2013).

$$min\,NRC = \sum\nolimits_{i=1}^{h} \left\{ \sum\nolimits_{k=1}^{n} \left[u_{i,k} \times \left(F\left(P_{i,k}\right) + C_k^u \times \left(1 - u_{i-1,k}\right)\right) + \left(1 - u_{i,k}\right) \times u_{i-1,k} \times C_k^d \right] \right\} \times D \tag{3.10}$$

Subject to:

$$\sum_{k=1}^{n} u_{i,k} \times P_{i,k} = P_i^L - P_i^R - P_i^S - P_i^V, \ \forall i \tag{3.11}$$

$$\sum_{k=1}^{n} u_{i,k} \times P_k^{max} \geq SR_i + P_i^L - P_i^R - P_i^S - P_i^V, \ \forall i \tag{3.12}$$

$$u_{i,k} \times P_k^{min} \leq P_{i,k} \leq u_{i,k} \times P_k^{max}, \ \forall i, k \tag{3.13}$$

$$u_{i,k} = 1\, for \sum_{i=i_{su}}^{i-1} u_{i,k} < MUT_k, \ \forall k \tag{3.14}$$

$$u_{i,k} = 0\, for \sum_{i=i_{du}}^{i-1} \left(1 - u_{i,k}\right) < MDT_k, \ \forall k \tag{3.15}$$

where,

NRC refers to the annual operation and maintenance cost for non-renewable energy units;
D is number of days of the target year;
P_i^R is the total renewable power output at period i (including wind, hydro, geothermal, biomass, etc.);

P_i^S is the power output from centralized storage (e.g. pumped hydro) at period i (which electricity comes from surplus renewable energy sources), according to a strategy for daily generation profile, tuned to supply electricity at off-valley periods (assuming a constant power output);
P_i^V is the power output from EVs V2G at period i, according to a strategy for daily V2G profile with a constant power output (included in the UCED as input data, disregarding the uncertainty inherent to the drivers/users' behavior);
P_i^L is the electricity load at period i, including additional load from EVs' charging, disregarding the uncertainty inherent to the drivers/users' behavior (the stored electricity by the centralized storage facility is not accounted in the load, since it is conceived as a service to the system, instead of a load to serve).

The previous objective functions are subject to five constraints, which are described as follows.

- System power balance demand: in each period, the committed units must satisfy the total load demand.
- Spinning reserve requirements: the spinning reserve is the unused capacity which can be activated by decision of the system operator. It is provided by devices that are synchronized to the network and able to affect the active power (Rebours and Kirschen 2005). A commonly used deterministic criterion sets the desired amount of spinning reserve so that the system can withstand the outage of any single generating unit without performing load shedding. This rulen is also known as N-1 criterion (Wood and Wollenberg 1996; Holttinen et al. 2012). Many system operators set the spinning reserve to be enough to handle the outage of the largest power unit in service (REN 2004; Vazquez 2006).
- Unit generation limits: thermal units are neither technically capable of producing below a given minimum production level, nor above a maximum.
- Unit minimum up and down times: if a unit is "on" it must be kept "on" for at least MUT periods of time. On the other hand, if a unit is "off", it must remain "off" for at least MDT periods of time.

3.2.4 Power Purchase Agreements and Feed-in Tariffs

The first movements towards competition in power generation have emerged following the 1970s oil crisis and as a response to lack of power generation investment from state-owned vertical integrated utilities. Let alone, energy security (from a geopolitical viewpoint) and adequacy (from a technical perspective) sparked the entry of new players into a system so far ruled by vertical integrated utilities. In 1978, the US put in place the Public Utilities Regulatory Policies Act (PURPA), which gave to Independent Power Producers (IPP)—mostly small-scale RES and cogeneration—the possibility to supply electricity to the incumbent utility. The latter was required to purchase power from these non-utility power generators (qualifying facilities) at the avoided cost, i.e. the cost the incumbent would incur either by generating or purchasing the corresponding electricity from another source. Rather recently, feed-in tariffs were designed to thrive renewable energy sources. In the early version of the feed-in mechanism, renewable generation (e.g. wind) was considered non-dispatchable energy, having priority over conventional units (e.g. coal and gas power plants). Feed-in generation has gained a significant expression in the EU, viz. in Germany, Spain and Portugal. Accordingly, PURPA and then feed-in tariff regime were the first tiers of competition in the power generation layer. However, PURPA did not cover the whole dynamic between the incumbent utility and IPP (Batle 2013).

In this context, power purchase agreements (PPA) emerged to govern the interparty relations and costs arrangements. The PPA prime goal is interparty risk sharing, standing as a legal guarantee to private investment. The PPA can be signed between the IPP and the incumbent utility, the system operator, a distribution concession, a supplier or even a large consumer (Harris 2006). Typically, a PPA is a long-term contract (more than 10 years of financial horizon), with fixed prices per unit of power (MW) and energy (MWh).

3.2.5 *Market-Based Model*

(i) Overview

Hogan (2001) stated "Good coordination cannot overcome bad market design. Markets in power, more than most markets, are made, they don't just happen". Indeed, on the contrary to other markets, the electricity is not a tradable commodity, independent of state investment and regulation.

The power system is a capital-intensive sector. On the one hand the transmission and distribution grids are regulated businesses (with cost recovery principles), on the other hand power generation depends of the cash-flow driven by energy trading. The energy market has different time horizons, i.e.: long-term; day-ahead; and intra-day.

From a chronological point of view, long-term markets operate before the day-ahead exchange or pool. That is to say, in the day before the target date, the players without previous bilateral contracts apply their offers and bids to the market operator. Afterwards, the TSO performs a technical assessment of the bilateral contracts and the day-ahead market results, in order to solve grid constraints and draw a final supply and demand program. Market agents are free to correct deviations before gate closure. In the EU, there is still an intra-day market managed by the exchange, in which agents submit additional supply offers and demand bids. Gate closure arises a few hours before real time. Following the market closure, the TSO ought to balance generation and load in real time (through the balancing market), considering the agents' upward and downward bids (Batle 2013) (Fig. 3.8).

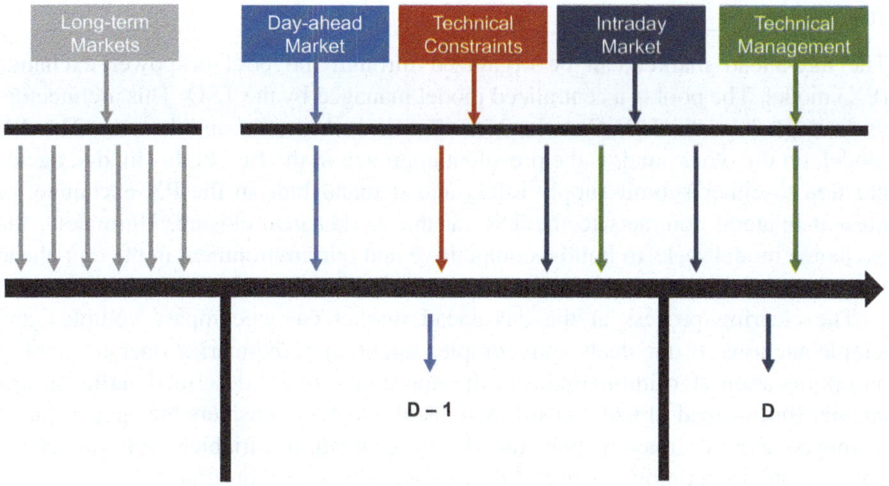

Fig. 3.8 Sequence of the wholesale market

(ii) Long-term Markets

As previously introduced, long-term markets provide to the agents a way to mitigate the short-term market risk. It is indeed a hedging strategy available to the wholesale market players. The contracts (usually lasting less than two years) are materialized either by financial or physical arrangements (which ensure the electricity supply in the absence of grid constraints).

These market instruments are traded through bilateral over-the-counter (OTC) contracts and futures markets. In the OTC market, supply and demand agents settle the power delivery terms, being therefore physical contracts. These forward contracts are negotiated outside an organized market. Regarding the derivatives instruments (pure financial products), wholesale agents can trade futures and options contracts. Futures are traded in organized exchanges in a standardized way (i.e. delivery date, location, quality and quantity), whereas option contracts consist in a list of buying (call) and selling (put) options for negotiation. Option contracts include: (i) plain vanilla options (where the owner have the right, but not the obligation, of buying a certain amount of energy at a predefined price established at the expiration date); (ii) spark spread options (it is an instrument for the evaluation of the thermal electric power plants through their efficiency rate in order to generate an electricity price); and (iii) swing options (in this case, the owner can buy a prearranged amount of electricity at a certain price and still have some flexibility in the amount of purchase energy and corresponding price) (Batle 2013; Deng and Oren 2006).

(iii) Day-ahead Markets

The day-ahead market can be organized through the pool or power exchange (PX) model. The pool is a centralized model managed by the TSO. This architecture was adopted in the US, Canada, New Zealand, Australia and Ireland. The PX model, on the other hand, is the prevalent approach in the EU. In this model, agents are free to either submit supply offers and demand bids in the PX exchange, or present bilateral contracts to the TSO at the market gate closure. Ultimately, the exchange model seeks to build a competitive and fair environment in the day-ahead wholesale market.

The clearing process at the day-ahead market can encompass complex and simple auctions. If one deals with complex auctioning, the market operator applies an optimization algorithm similar to the unit commitment described earlier in this chapter (but instead of using cost functions, this model considers the agents' bids). Complex auctions may include the power generation variables that impact the power plant's operation, such as the minimum down and up times.

Alternatively, in the simple auctioning agents submit offers or bids characterized by the price and quantity of electric energy to buy or sell. Then, a merit order is established by sorting bids in descending order and ranking offers in ascending order. The equilibrium market price is found by the interception of supply and demand curves. The equilibrium price is thus the bid price for the amount of energy that corresponds to the cumulative amount of energy demanded. The supply offers with prices equal or below to the market clearing price are accepted. On the other hand, the demand bids with prices not lower than the equilibrium price are accepted (Batle 2013) (Fig. 3.9).

There is still another auctioning possibility, emerging as a strategy to tackle lack of offering transparency through simple auctions, i.e.: semi-complex auctions. Since simple auctions do not consider the dynamics embedded in the power generation unit commitment, thermal power plants (with start-up and shut-down constraints) could include, in the selling offers, an extra cost to curb uneconomical scenarios. Semi-complex auctions arise as a response to this drawback and are characterized by including just a few technical constraints of the generating units or allowing inter-temporal constraints in the offers. In the EU power exchange, the latter option is available in the form of block-bids (Batle 2013).

(iv) Intraday Markets

Adjustment markets were drawn to provide, to the agents, the capability of fine-tuning the day-ahead market schedule, through additional energy offers and bids, after day-ahead market closure.

One can find two alternative models for adjustment markets. In the EU, the intraday market is settled via power exchange and the balancing market is managed by the TSO, while in the US there is the real-time market (centralized) to balance deviations between the day-ahead requirements for electricity and actual real-time demand levels.

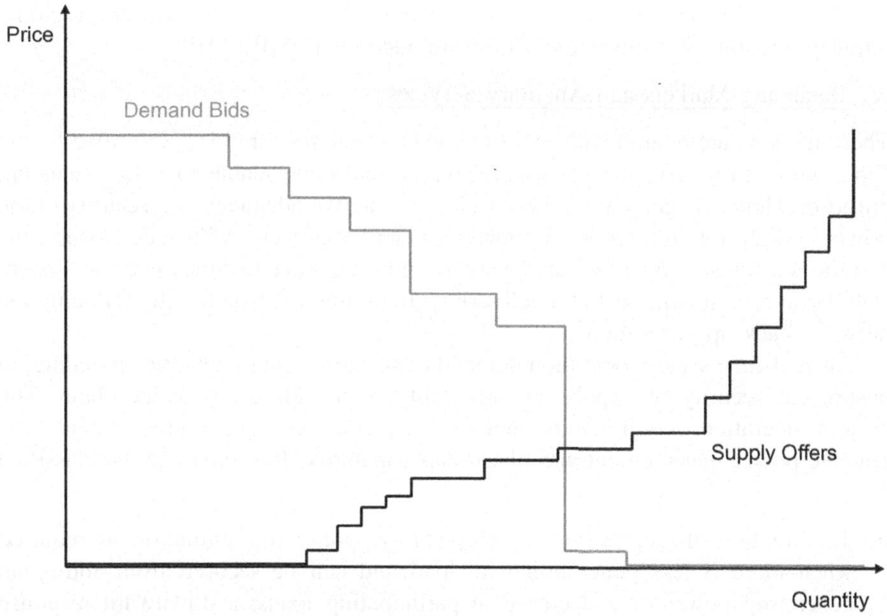

Fig. 3.9 Theoretical day-ahead market clearing

In the Iberian intraday market there are at least six day-ahead sessions of the intraday auctions market, with different programming horizons for each session. This market manages the price areas of Portugal and Spain, and the free capacity of the interconnections: Spain-Portugal, Spain-Morocco and Spain-Andorra. This market is organized as follows:

- Once the feasible day-ahead schedule is set, the intraday auction session begins for the periods included in the day-ahead program;
- Each intraday auction session can have the purpose of one or several schedule periods;
- Each schedule period may be subject to successive intraday market auction sessions;
- Schedule periods for which no feasible day-ahead schedule exists are not included in the intraday market auctions (OMIE 2018).

Recent EU regulation established a guideline on capacity allocation and congestion management. This initiative has set up an objective model for intraday markets based on the continuous negotiation of energy in the market. The continuous intraday market benefits the agents' imbalances management, as they can make use not only the liquidity of the national market, as well as the liquidity in the other bidding zones. This market aims at easing energy trade between different bidding zones of Europe and increasing the efficiency in intraday markets. The intraday continuous market is deemed compatible with both current regional intraday auctions and future pan-European auctions (OMIE 2019).

(v) Balancing Markets and Ancillary Services

These markets are related with real-time power system operation, performed by the TSO, who can modify the generators' output and controllable load for balancing purposes. Hence, balancing markets seek to ensure the adequacy and security of the power system, through an economically efficient approach. After gate closure, the TSO urges for supply offers and demand bids, in order to guarantee the system stability at a minimum cost and delivering environmental benefits (by reducing the need for back-up generation).

The real-time system operation demands a systemic and systematic procedure, to ensure the security of supply and reliability of the electricity value chain. The system operation requirements include frequency control, voltage regulation, reactive power management and black-start capability. Reserves may be classified by:

- the direction of their actions—an upward response (up-regulation) is required when there is less generation than load and can be secured from additional generating power or a decrease in participating loads; a downward response (down-regulation) is required when there is more generation than load and can be obtained from a drop in generating power or an increase in participating loads;
- the required response time—fast response in order to arrest a frequency drop; slower response for reserve that replaces other reserve categories (a matter of minutes or tens of minutes); for longer time scales, there can be also reserve provision for additional reserves, to counter forecast errors.

Although the universality of reserve constraints, the technical procedures could vary from country to country. For instance, regulation reserve is referred in the US as fast response to normal variations in load or generation, both upward and downward. Alternatively, the European Network of Transmission System Operators for Electricity (ENTSO-E) considers three categories of reserves, i.e.: primary, secondary and tertiary control (Holttinen et al. 2012) (Fig. 3.10).

A proposal for assessing the performance of the operating reserve was presented in Lopes et al. (2008) and Matos et al. (2009), through the RESERVAS model. According to this model, at a certain period of time, minimum number of units will have to be dispatched to satisfy the forecasted load and the specified needs for primary and secondary reserve. In order to complete the operating reserve, units that could be available in less than one hour must be identified (tertiary reserve) (Lopes et al. 2008).

In this model, the power balance equation is set to assess the risk indices associated with the operating reserve.

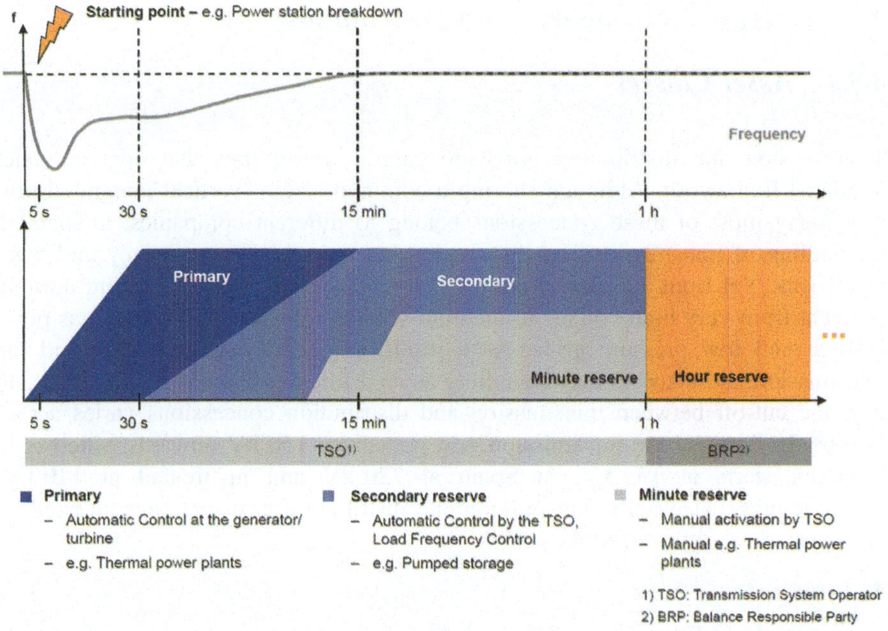

Fig. 3.10 Chronology of frequency control (RWE 2008)

$$R_{OPE} = R_S + R_T < \Delta L + \Delta P_W + \Delta G \tag{3.16}$$

where,

R_{OPE} refers to the operating reserve at period t;
R_S is the secondary reserve at period t;
R_T is the tertiary reserve at period t;
ΔL stands for the short term load deviation at period t;
ΔP_W refers to the possible wind power capacity variation at period t;
ΔG represents the generating capacity variation due to forced outages at period t.

The previous equation describes the risk of changes in the load, wind power capacity and generating outages not being properly covered by the amount of spinning reserve, and also by those generators that can be synchronized within 1 h.

From a market point of view, the TSO can purchase ancillary services before and after gate closure, in short-term markets. In the ancillary services market, generation companies offer frequency control, subject to a payment from the TSO. Primary reserve is off-market in several countries (where it is mandatory for generation units), whereas secondary and tertiary reserves are often provided in a market environment.

3.3 Energy Transmission and Distribution

3.3.1 Asset Classes

Transmission and distribution grids are natural monopolies that operate under regulated framework. Although sharing a common origin (vertical integrated utility), today most of these concessions belong to different companies. Ownership unbundling of transmission and distribution has enhanced transparency and business focus. Yet from a technical point of view, the "wires" and "rust" are not that different from very high voltage to medium voltage. The same applies to gas pipes in high and low pressure grids. As a result, engineers tend to understand the transmission and distribution unbundling as an artificial separation of the grids. In fact, the cut-off between transmission and distribution concessions varies across Europe. In Portugal the transmission grid goes from 150 kV onwards, whereas in Denmark starts at 132 kV, in Spain at 220 kV and in Ireland at 110 kV. Unsurprisingly, electricity transmission and distribution grids are characterized by the same set of main asset classes.

- Lines:
  - Overhead lines;
  - Underground lines.
- Substations:
  - Power transformers and autotransformers;
  - Circuit-breakers;
  - Shunt reactors;
  - Capacitor banks;
  - Instrument transformers;
  - Disconnectors;
  - Surge arresters.
- Systems:
  - SCADA;
  - Protection, automation and control systems;
  - Metering;
  - Monitoring systems;
  - Telecommunication infrastructure.

In this book, asset management of natural gas systems does not enclose the upstream layer. One focuses the analysis on the LNG regasification, transmission and distribution grids, gathering the next main installations.

- LNG regasification terminal.
- Pipes:
  - onshore (overland) pipes;
  - offshore (subsea) pipes;
  - inshore water crossings.
- Stations:
  - gas regulation and metering stations;
  - block valve stations;
  - custody transfer stations;
  - interconnection stations;
  - junction stations.
- Underground storage.

Amongst natural gas infrastructure, there is a subset of asset classes that are worth mentioning (in view of asset management priorities):

- Regulators;
- Compressors;
- Boilers;
- Gas odorization systems;
- Control systems.

3.3.2 Regulation Models

Energy transmission and distribution grids are regulated businesses, being envisaged as natural monopolies. The two main regulation models for natural monopolies are the cost-of-service (also called rate-of-return) and the incentive-based approach. In the cost-of-service model, the concessionaire remuneration (rate level) is carried out through the next logic:

(i) Identification of the company's total costs and its business plan;
(ii) Definition of the rate-of-return to ensure the capital cost recovery.

The concessionaire's regulated revenues are computed via Eq. (3.17) (Gómez 2013).

$$AR = TC = O\&M + DP + RoR \times RAB + TAX - ADR \tag{3.17}$$

where,

AR refers to allowed revenues;
TC refers to total cost of service;
O&M refers to operation and maintenance costs;

DP refers to depreciation expenses;
RoR refers to rate-of-return;
RAB refers to regulatory asset base;
TAX refers to taxes;
ADR refers to additional revenue.

There are two critical moments for a grid company, when it comes to economic regulation: (i) the process of assessment, discussion and eventually approval/disapproval of the multi-year grid investment plan (which will integrate the RAB); and (ii) the announcement of the regulatory model and parameters (especially the RoR) for the subsequent years.

The RAB consists of the net value of the company's assets, being yearly updated by adding the new investment projects (i.e. CAPEX) transferred to exploration and subtracting depreciations. Concerning the RoR, the weighted average cost of capital (WACC) is the most popular method for computing this rate. The WACC ponders the different financing sources of the company.

The cost-of-service model encloses a drawback, referred as Averch-Johnson effect. This negative externality emerges if the utility's RoR is higher than the real WACC. In this context, the company is incentivized to overinvest, leading to an economic inefficiency. On the other hand, when the grid concessionaire's RoR is lower than the WACC the company will have no motivation to invest. This underinvestment scenario, besides jeopardizing the grid's reliability, also results in an economic inefficiency. The Averch-Johnson effect arises due to the information asymmetry between the regulatory authority and the grid concessionaire. This information asymmetry is hard to tackle—owing to the time lag between the utility's technical and financial moves, and the regulator's thorough awareness—being worsened in the cost-of-service model.

Despite the presented drawback, the cost-of-service is still a robust approach, since: (i) it provides financial stability to the regulated company; (ii) the RoR is set by the regulator, who holds the capability of fine-tuning the rate to an efficient level (for cost recovery); and (iii) the regulator and the concession granter (the government) are able to balance the proposed CAPEX (by the utility) with the necessary grid investments (e.g. for reliability, RES integration and interconnection purposes), through a selective approval of investment projects (via multicriteria/cost-benefit analysis) and a cost-effective RoR (that prevents overinvestment).

When it comes to the incentive-based model, the regulatory periods are longer (4–5 years) to provide improvement opportunities (mobilized by incentives). This model is deemed as a midpoint between the deregulated environment (with market-based prices) and the cost-of-service approach (Gómez 2013; CEER 2019). The incentive-based approach is built upon three main principles: (i) the reward/penalty structure; (ii) giving the company freedom to choose its goals; and (iii) providing the operator latitude in how it will achieve its goals. The incentive-based model can be implemented through the price cap and revenue cap methods.

Price cap regulation allows the operator to change its price level according to an index *I* (usually related with inflation) and a productivity offset (the *X-factor*). In this regulation model, the national regulation authority groups the operator's activities into price or service baskets and, then, sets an $I - X$ *index* (the price cap index for each basket). According to the established price baskets, the grid operator is able to change prices within the basket, providing that the average variation of the service prices in the basket does not exceed the price cap index. Concerning the revenue cap, this model is similar to price cap regulation, but instead of considering the price cap index, there is the establishment of a revenue cap index. Revenue cap seems rather adequate when costs do not vary substantially with units of sales (Jamison 2007).

According to CEER (2019), efficiency requirements are an effective strategy to curb the grid operators' costs, without undermining the quality of service. In Europe, most national regulatory authorities set efficiency targets, frequently focused on OPEX of power and gas transmission and distribution. CAPEX efficiency of energy grids, though, is addressed at a limited scale. Finally, TOTEX (total expenditures) efficiency requirements are tackled in a few cases (CEER 2019) (Table 3.2).

In Europe, the incentive-based approach is the prevalent regulation model. Although the revenue cap outranks (in number) the price cap method, in most cases one observes a blended approach of a cap regulation (revenue or price) and a rate of return (Table 3.3).

The incentive-based regulation brings clear benefits in terms of continual improvement and efficiency requirements, besides mitigating information asymmetries between the regulator and the concessionaires. This type of regulation is also helpful in tackling overinvestment. Yet in some cases, service quality and security of supply can be jeopardized (Jamison 2007).

3.3.3 Regulatory Asset Base

In natural monopolies, the RAB consists in the cornerstone of the utility's remuneration. In a simplified way, the RAB stands for the total non-depreciated asset value. This is the valued to be multiplied by the RoR to compute the capital remuneration. Therefore, the RAB is a critical variable for the utility's annual cash-flow.

The RAB should be formed by the assets necessary for the provision of the regulated business and encloses several components, such as fixed assets, working capital or construction in progress. The most applied methods to determine the RAB are the: book value; reproduction cost; replacement cost; and market value.

The new assets installed in the grid are included in the RAB through an *ex ante* methodology of approved CAPEX. Yet this approach can lead to information asymmetry between the concessionaire and the regulator. In this case, two situations could arise during the regulatory period:

Table 3.2 Efficiency requirements on CAPEX and OPEX, in European energy grids [adapted from CEER (2019)]

Country	Electricity				Natural Gas			
	Efficiency requirement applied on the CAPEX?		Efficiency requirement applied on the OPEX?		Efficiency requirement applied on the CAPEX?		Efficiency requirement applied on the OPEX?	
	TSO	DSO	TSO	DSO	TSO	DSO	TSO	DSO
Austria	No	Yes	Yes	Yes	No	Yes	Partially	Yes
Belgium	No	Depends of region	No	Depends of region	No	Depends of region	No	Depends of region
Czech Republic	No	No	Yes	Yes	No	No	Yes	Yes
Germany	Yes	Yes	Yes	Yes	Yes	Yes	Yes	Yes
Estonia	No	No	No	No	No	No	No	No
Spain	No	No	No	No	Yes	No	No	No
Finland	No	No	Yes	Yes	No	No	Yes	No
France	No	No	Yes	Yes	No	No	Yes	Yes
Great Britain	Yes	Yes	Yes	Yes	Yes	Yes	Yes	Yes
Greece	No	No	No	No	No	No	No	No
Hungary	No	No	Yes	Yes	No	No	Yes	No
Ireland	Yes	Yes	Yes	Yes	No	No	Yes	Yes
Italy	No	No	Yes	Yes	No	No	Yes	Yes
Lithuania	No	No	Yes	Yes	No	No	Yes	Yes
Luxembourg	Partially	Partially	Yes	Yes	Partially	Partially	Yes	Yes
Latvia	No	No	No	No	No	No	No	No
Netherlands	Yes	Yes	Yes	Yes	Yes	Yes	Yes	Yes
Norway	Yes	Yes	Yes	Yes	N.I.	Yes	N.I.	Yes
Poland	No	No	No	Yes	No	No	No	No
Portugal	Yes	Yes	Yes	Yes	No	No	Yes	Yes
Romania	No	No	Yes	Yes	No	No	Yes	Yes
Sweden	No	No	Yes	Yes	No	No	Yes	Yes
Slovenia	No	No	Yes	Yes	No	No	Yes	Yes

Legend N.I.: no information

- The utility invests less than what has been previously estimated in the RAB, resulting in higher profits;
- The utility invests more than the approved plan, leading to financial losses.

In view of this drawback, most regulators adopt a blended *ex ante*/*ex post* approach, where the authorized investments in the *ex post* review replace the ones included *ex ante* in the RAB. In the *ex post* model, the larger the period between the investment inclusion in the RAB and the review, the higher will be the incentive for actual capital cost cut (Gómez 2013; Jamison 2007).

Table 3.3 Regulation models in European energy grids [adapted from CEER (2019)]

Country	Electricity		Natural Gas	
	TSO	DSO	TSO	DSO
Austria	Rate-of-return	Price cap	Combined model of price cap (opex) and rate-of-return (capex)	Price cap
Belgium	Revenue cap + cost control incentives + quality related incentives	Depends of region	Revenue cap + cost control incentives	Depends of region
Czech Republic	Revenue cap	Revenue cap	Revenue cap	Revenue cap
Germany	Revenue cap	Revenue cap	Revenue cap—incentive based	Revenue cap
Estonia	Rate-of-return	Rate-of-return	Rate-of-return	Rate-of-return
Spain	Rate-of-return + incentive based regulation	Rate-of-return + incentive based regulation	Revenue cap	Revenue cap
Finland	Revenue cap	Revenue cap	Revenue cap	Revenue cap
France	Revenue cap, incentive based with pass through	Revenue cap, incentive based with pass through	Revenue cap, incentive based with pass through	Revenue cap, incentive based with pass through
Great Britain	Revenue cap based on rate-of-return with incentive-based regulation	Revenue cap based on rate-of-return with incentive-based regulation	Revenue cap based on rate-of-return with incentive-based regulation	Revenue cap based on rate-of-return with incentive-based regulation
Greece	Revenue cap	Cost plus	Cost plus	Revenue cap
Hungary	Combined model of incentive-based regulation and cost-plus	Incentive-based regulation (mixture of price cap, revenue cap and quality regulation)	Incentive-based regulation (mixture of price cap, revenue cap and quality regulation)	Incentive-based regulation (mixture of price cap, revenue cap and quality regulation)

(continued)

Table 3.3 (continued)

Country	Electricity		Natural Gas	
	TSO	DSO	TSO	DSO
Ireland	Revenue cap based on rate-of-return with incentive-based regulation	Revenue cap	Revenue cap based on rate-of-return with incentive-based regulation	Revenue cap based on rate-of-return with incentive-based regulation
Italy	Combined model of price cap (opex) and rate-of-return (capex)	Combined model of price cap (opex) and rate-of-return (capex)	Combined model of price cap (opex) and rate-of-return (capex)	Combined model of price cap (opex) and rate-of-return (capex)
Lithuania	Price cap	Price cap	Price cap	Price cap
Luxembourg	Revenue cap	Revenue cap	Revenue cap	Revenue cap
Latvia	Cost-plus/rate-of-return	Cost-plus/rate-of-return	Cost-plus/rate-of-return	Cost-plus/rate of return
Netherlands	revenue cap	Price cap	Revenue cap	Price cap
Norway	Revenue cap—incentive based	Revenue cap—incentive based		
Poland	Cost of service (with elements of revenue cap)	Mixed (revenue cap with elements of incentive-based regulation) with elements of quality regulation	Cost of service (with elements of revenue cap)	Cost of service (with elements of revenue cap)
Portugal	Combined model of price cap (opex), standard costs in new investments and rate-of-return (capex)	HV/MV: combined model of price cap (opex) and rate-of-return (capex); LV: totex	Combined model of price cap (opex) and rate-of-return (capex)	Combined model of price cap (opex) and rate of return (capex)
Romania	Revenue cap	Price cap for DSO with concession contract	Revenue cap	Price cap
Sweden	Revenue cap	Revenue cap	Revenue cap	Revenue cap
Slovenia	Revenue cap	Revenue cap	Revenue cap	Revenue cap

3.3.4 Capital Cost

The theoretical model of the cost of capital is built on the Capital Asset Pricing Model (CAPM). The CAPM provides an explanation of the relationship between risk and asset returns, being developed through the Markowitz mean-variance-efficiency model. In the CAPM, risk-averse investors are oriented towards expected returns and the variance of returns (risk). Therefore, the financial agents will invest in an efficient portfolio, which maximizes the expected results for a given risk level, in a coherent way with the risk aversion degree of each agent. Accordingly, taking into account the risk aversion degree, the agent's portfolio will enclose risky and risk-free assets, in a different proportion.

The risk-free nominal rate—made up by the inflation rate and the risk-free actual tax rate—and the market expected profitability are independent of the asset risk. They are indeed attributes of the markets where the securities are traded and concerned with each asset (Gómez 2013; Eugene and Fama 2004).

According to the CAPM, the weighted average cost of capital (nominal, pre-tax) is computed as presented in Eq. (3.18) (Bishop and Officer 2013; ERSE 2017).

$$WACC_{(nominal,\,pre-tax)} = \frac{R_{Equity}}{1-T} \times \frac{Equity}{Equity + Debt} + R_{Debt} \times \frac{Debt}{Equity + Debt} \tag{3.18}$$

where,

R_{Equity} refers to cost of equity;
R_{Debt} refers to cost of debt;
Equity refers to market value of the firm's equity;
Debt refers to market value of the firm's debt;
T refers to tax rate.

In which,

$$R_{Equity} = R_f + \beta \underbrace{[R_m - R_f]}_{\text{Market risk premium}} \tag{3.19}$$

where,

- R_f—nominal *risk-free rate of return*;
- β—equity beta (i.e. the risk of a given asset relative to the risk of the market);
- R_m—expected annual return of the market portfolio.

The R_{Debt} is calculated by adding the R_f to the debt risk premium (*DRP*).

$$R_{Debt} = R_f + DRP \tag{3.20}$$

As alternative to the pre-tax WACC, there is the "plain vanilla" WACC which assumes all tax effects of financing.

3.3.5 Operation Cost

Electricity and gas grids, besides capital costs, have to handle operation expenditures in order to ensure a reliable energy transmission and distribution. OPEX budgeting is one of the most important asset management competencies. Ideally, utilities draw a multi-year OPEX plan (linked with the regulatory period), which is annually reviewed and subject to the company's executive committee approval. The OPEX budget includes the next items (non-exhaustive) (Fig. 3.11):

- Human resources;
- Preventive and corrective maintenance actions;
- Asset inspections;
- Vegetation management;
- Environment, safety and surveillance procedures;
- Information and communication systems;
- Taxes;
- Insurance costs.

The most meaningful OPEX are related with human resources and asset inspection and maintenance, which together can sum up roughly 80% of the total operation costs. Usually, workforce and asset-related outsourcing are evenly weighed in the OPEX.

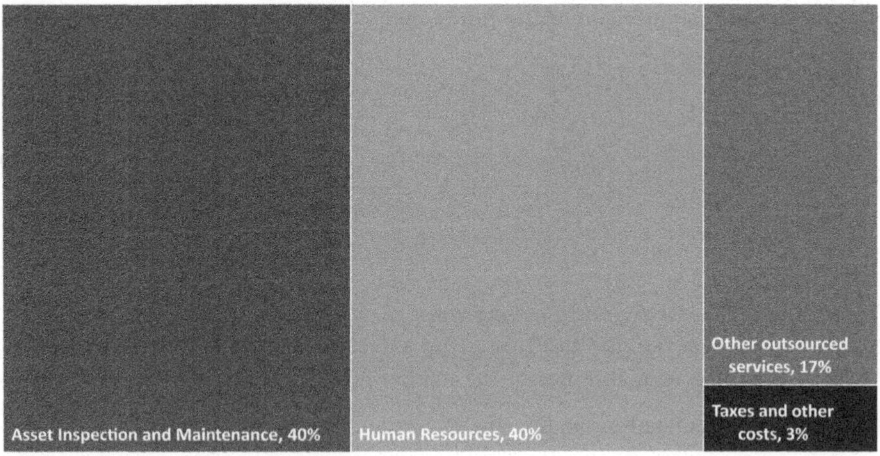

Fig. 3.11 Approximate OPEX breakdown in energy grids

3.4 Liquefied Natural Gas

3.4.1 Asset Classes

Liquefied natural gas is composed of methane, which has been converted to liquid form. Since LNG stands for just 1/600 of the natural gas volume, the liquefied form of gas brings clear advantages when it comes to its transportation. The liquefaction process consists of cooling natural gas to −162 °C until it turns out a liquid. LNG is then converted back to gas, in regasification terminals, for gas transmission, distribution and end use.

Although LNG regasification is embodied in the natural gas midstream, it differs significantly from transmission and distribution grids. An LNG terminal is rather an industrial operation, with a wide set of processes and asset classes (Fig. 3.12).

From an asset management perspective, the most important asset classes in an LNG receiving terminal are listed next:

- Unloading arms;
- Cryogenic pipelines;
- Storage tank(s);
- Low pressure (LP) pumps;
- Boil-Off Gas (BOG) compressors and recondensers;
- High pressure (HP) pumps;

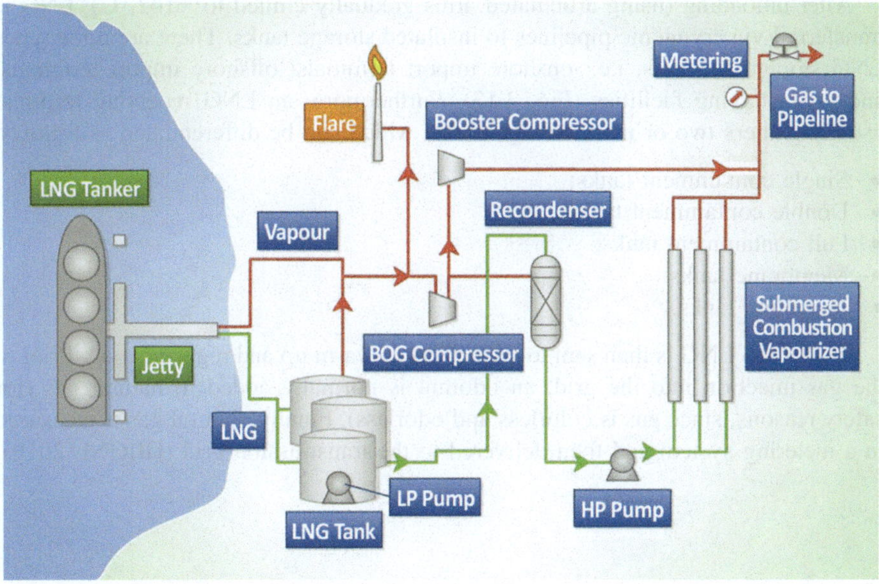

Fig. 3.12 LNG receiving terminal process (Canaport 2019)

Fig. 3.13 Example of an LNG terminal infrastructure (REN 2019)

- Vaporizers;
- Gas metering;
- Command and Control Systems.

After unloading (using articulated arms gradually chilled to −162 °C), LNG is transferred via cryogenic pipelines to insulated storage tanks. There are three types LNG storage facilities, i.e.: onshore import terminals; offshore import terminals; and peak-shaving facilities (Fig. 3.13). Furthermore, an LNG receiving terminal usually gathers two or more storage tanks, which can be differentiated as follows:

- Single containment tanks;
- Double containment tanks;
- Full containment tanks;
- Membrane tanks;
- In-ground tanks.

The stored LNG is then sent to vaporizers to warm up and regasification. Prior to the gas injection into the grid, an odorant is normally added to natural gas (for safety reasons, since gas is colorless and odorless). Finally, natural gas is measured in a metering system and then delivered to the transmission grid (GIIGNL 2019).

3.4.2 Business Model

The business model of LNG terminals differs from natural monopolies, like transmission and distribution grids. In fact, LNG plants compete in the international arena. In this section a brief outline is provided on the services carried out by LNG terminals.

In the EU, the majority of the terminals have a Regulated Third-Party Access Regime (TPA) in place, comparing to a limited number of plants with negotiated access conditions (Dunkerque, Gate and the three UK terminals) and another terminal with an "hybrid" access regime (Porto Levante) (Fig. 3.14).

The service portfolio in the different EU LNG terminals is somehow dependent of the gas historical framework in each country. Natural gas industry (like the energy sector as a whole) is strongly shaped by national public policies and regulation, such as power generation investment, capacity interconnection, market competition and energy independence. This context leads to a heterogeneous LNG ecosystem in the EU. Still one can find out two main types of services, i.e.: bundled and unbundled services.

The common ground for bundled services in the EU refers to "ship unloading + LNG storage + regasification (send-out)" service. This bundled service gathers the "bread and butter" of an LNG terminal. That is to say, the LNG ships berth and unload, the LNG is converted back into natural gas, being then injected into the transmission grid. A variant approach for bundled service consists of "ship unloading + LNG storage + trucks loading". Besides the typical bundled service (unloading + storage + regasification) the majority of the LNG plants also provide unbundled services, in order to enhance flexibility of the bundled service or simply to supply additional and independent services.

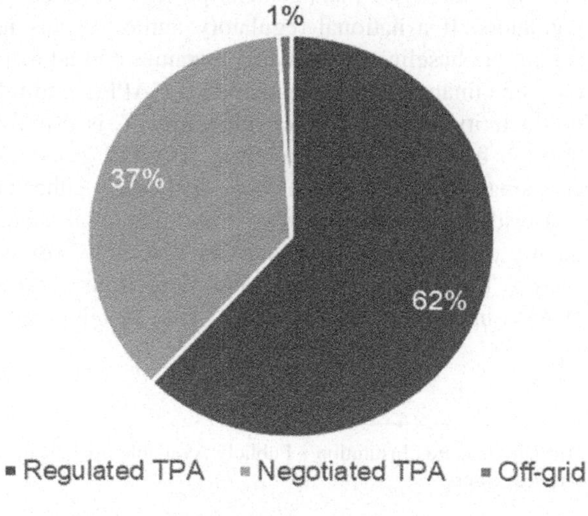

Fig. 3.14 Access regimes in place to LNG send-out capacity in EU terminals (% vs total send-out capacity) (CEER 2017)

Bearing in mind market competition and integration drivers, CEER (2017) urged national regulatory authorities and LNG system operators to improve access to the terminals and to foster the use of these infrastructures and liquidity. Actually, transparency is key to thrive the usage of LNG terminals and the number of newcomers. Additionally, CEER (2017) highlights the lack of service standardization, owing to each country's idiosyncrasies.

To conclude, LNG services should be more flexible—to increase market competition—transparent and dynamic (by providing innovative services), road to the development of European LNG hubs.

3.5 Future Regulatory Perspectives

Considering the energy industry trend (addressed in Chap. 2), one can expect regulatory changes in the power and gas value chains. At a European level, there will likely be an intensification of the *regionalization* and *europeanisation* of grid planning and system and market operation. At a national level, TSO and DSO interaction will have to be reinforced, both from a planning and operational perspective. Furthermore, the mission and services carried out by electricity TSOs and DSOs will probably evolve, due to the massive RES penetration in distribution grids; active grid management; new market services, linked with electric mobility and demand response; and the increasing role of local energy markets, smart cities and internet of things. Accordingly, the legal and regulatory frameworks must be adapted, to allocate each operational function and consider the concerned costs and liabilities (Glachant et al. 2015).

Concerning natural monopolies' regulation, TSOs and DSOs may be progressively subject to the TOTEX approach, as a consequence of the asset management driver. In fact, several utilities have been adopting PAS 55[1] or ISO 55000 to comply with asset management principles, directly or indirectly prescribed by regulators. If a national regulatory authority puts in place a TOTEX model for setting the baseline of regulated revenues and adjustment parameters, there will be only one financial envelope that adds CAPEX and OPEX. In this case, the utility's productivity factor for efficiency gains, X, is usually established by the regulator through benchmarking. Therefore, TOTEX gives the utility higher flexibility for choosing—via tradeoff analysis—cost cut in either OPEX or CAPEX.

Besides the energy grids' primordial goal concerned with the security and quality of service, the regulator can also draw key performance indicators (KPIs), such as the consumers' satisfaction, environmental impact reduction, energy efficiency initiatives and research and development investment. This type of

[1]British Standards Institution's Publicly Available Specification for the optimized management of physical assets.

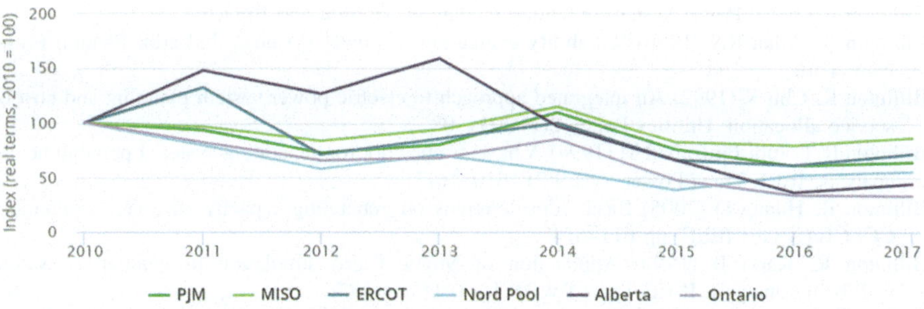

Fig. 3.15 Average wholesale electricity prices in the period 2010–2017 (IEA 2018)

mechanism is called performance-based regulation (PBR) and links the company's remuneration to the KPIs' degree of accomplishment (Gómez 2013).

In the UK, OFGEM (2010) has developed the RIIO model, which stands for *Revenue = Incentives + Innovation + Outputs*. This methodology sketches a new regulatory architecture to oversee power and gas natural monopolies. RIIO aims at encouraging energy grids to:

- place stakeholders in the center of the decision-making process;
- deploy the required investment projects to ensure adequate reliability levels;
- develop innovative approaches to optimize current and future activity costs;
- play a central role in the energy systems decarbonization.

Regarding electricity wholesale markets, since 2010 energy prices have dropped in some countries due to a plateau in demand and higher penetration of RES with low marginal costs (Fig. 3.15). IEA (2018) expects that in the medium to long term many markets will keep experiencing price pressure in wholesale energy market, as a result of additional RES integration (with near zero marginal cost). This phenomenon together with negative wholesale energy prices (e.g. in Germany), could accelerate a market redesign that can support the energy transition while ensuring the security and adequacy of power systems.

References

Allan RN, Takieddine FN (1977) Generation modelling in power system reliability evaluation, pp 47–50. IEEE Conference Publication, p 148

Allan RN, Silva AMLd, Abu-Nasser A, Burchett RC (1981) Discrete convolution in power system reliability, pp 452–456. IEEE Transactions on Reliability, R-30

Batle C (2013) Electricity generation and wholesale markets. In: Regulation of the power sector, power systems, 61. Editor Pérez Arriaga, I. J., Springer, Berlin

Bhavaraju MP (1974) An analysis of generating capacity reserve requirements, Paper no. C-74-155-0. IEEE

Billinton R (1966) Bibliography on application of probability methods in the evaluation of generating capacity requirements, Paper no. 31 CP 66–62. IEEE Winter Power Meeting

Billinton R (1970) Power system reliability evaliation. Gordon and Breach
Billinton R, Allan RN (1996) Reliability evaluation of power systems, 2nd edn. Plenum Press, New York
Billinton R, Chu K (1992) An integrated approach to electric power system planning and cost of service allocation. Utilities Policy 2(4):303–308
Billinton R, Fotuhi-Firuzabad M (1994) A basic framework for generating system operating health analysis. IEEE Trans Power Syst 9(3):1610–1617
Billinton R, Huang D (2005) Basic considerations on generating capacity adequacy evaluation. CCECE/CCGEI IEEE, pp 611–614
Billinton R, Karki R (1999) Application of Monte Carlo simulation to generating system well-being analysis. IEEE Trans PWRS 14(3):1172–1177
Bishop S, Officer RR (2013) Review of debt risk premium and market risk premium. Value Adviser Associates
Canaport LNG (2019) Available: https://www.canaportlng.com/lng-process. Accessed 25 July 2019
Carr M (2018) After topping commodity gains in 2018. Carbon set to surge again, Bloomberg, 27 Dec 2018
CEER—Council of European Energy Regulators (2017) Removing LNG barriers on gas markets, Ref: C17-LNG-32-03, Brussels
CEER—Council of European Energy Regulators (2019) Report on regulatory frameworks for European energy networks. Incentive Regulation and Benchmarking Work Stream
Decker C (2014) Modern economic regulation: an introduction to theory and practice. Cambridge University Press, Cambridge
Deng SJ, Oren SS (2006) Electricity derivatives and risk management. Energy 31(6):940–953
ERSE—Entidade Reguladora dos Serviços Energéticos (2017) Portuguese energy regulatory framework for TSO allowed revenues, Lisbon
Eugene KRF, Fama F (2004) The capital asset pricing model: theory and evidence. J Econ Perspect 18(3):25–46
European Commission (2010) Emissions trading: questions and answers on the EU ETS Auctioning Regulation, MEMO/10/338
European Commission—DG Climate Action (2011) General guidance to the allocation methodology
Farrer TH (1902) The State and its relation to Trade. Macmillan, London
GIIGNL (2019) The international group of liquefied natural gas importers. Information Paper No. 2, LNG Process Chain, Neuilly-sur-Seine
Glachant JM, Vasconcelos J, Rious V (2015) A conceptual framework for the evolution of the operation and regulation of electricity transmission systems towards a decarbonised and increasingly integrated electricity system in the EU. Florence School of Regulation, European University Institute, Florence
Gómez T (2013) Monopoly regulation. In: Pérez Arriaga IJ(ed) Regulation of the power sector, power systems, vol 61. Springer, Berlin
Harris C (2006) Electricity markets: pricing, structures and economics. Wiley, Hoboken
Hogan WH (2001) Federal energy regulatory commission technical conference on interregional coordination, Washington DC
Holttinen H, Milligan M, Ela E, Menemenlis N, Dobschinski J, Rawn B, Bessa RJ, Flynn D, Gómez-Lázaro E, Detlefsen NK (2012) Methodologies to determine operating reserves due to increased wind power. IEEE Trans Sustain Energy 3(4):713–722
Huang K, Yang H, Huang C (1997) A new thermal unit commitment approach using constraint logic programming. IEEE
ICE—International Exchange (2019) Available: https://www.theice.com/marketdata/reports/149. Accessed 28 June 2019
IEA—International Energy Agency (2018) World Energy Outlook, Paris
Jamison MA (2007) Regulation: price cap and revenue cap. In: Capehart BL (ed) Encyclopedia of energy engineering and technology, vol 3. CRC Press, New York, pp 1245–1251

Joscow PL (2008) Lessons learned from electricity market liberalization. The Energy Journal, no. Special Issue. The Future of Electricity: Papers in Honor of David Newbery, pp 9–42
Khatib H (2003) Economic evaluation of projects in the Electricity Supply Industry, London. The Institution of Engineering and Technology, United Kingdom
Leite da Silva A, Resende L, Manso L, Billinton R (2004) Well-being analysis for composite generation and transmission systems. IEEE Trans Power Syst 19(4):1763–1770
Leite da Silva A, Manso L, Sales W, Resende L, Aguiar M, Matos Mea (2007) Application of Monte Carlo simulation to generating system well-being analysis considering renewable sources. Eur Trans Electr Power 17(4):387–400
Li C, Johnson R, Svoboda A (1997) A new unit commitment method. IEEE Transactions
Lopes JP, Matos M, Cabral PG, Ferreira MS, Martins NF, Wert CA, Martos FS, Sanz RL, Rosa M, Ferreira R, Silva ALd, Sales W, Resende L, Manso L (2008) Dealing with intermittent generation in the long-term evaluation of system adequacy and operational reserve requirements in the Iberian Peninsula. CIGRE, vol C1-304
Martins N, Santos J, Damas M (2012) Integrating renewable energy sources in the Portuguese power system: new generation adequacy indicators. In 11th international workshop on large integration of wind power into power systems as well as transmission networks for offshore wind power plants, Lisbon
Matos MA (2007) Introdução ao problema de escalonamento e pré-despacho. Faculty of Engineering, University of Porto
Matos P (2011) O Leilão em Portugal: as escolhas dos operadores energéticos, EDP
Matos M, Lopes JP, Rosa M, Ferreira R, Silva ALd, Sales W, Resende L, Manso L, Cabral P, Ferreira M, Martins N, Artaiz C, Soto F, López R (2009) Probabilistic evaluation of reserve requirements of generating systems with renewable power sources: The Portuguese and Spanish cases. Electr Power Energy Syst 31:562–569
Mendonça A, Matos M, Lopes JP (2004) A multicriteria approach to identify the adequate wind power penetration in isolated grids. In: Proceedings of MedPower'04
Moreira da Silva M (2013) Energy planning with electricity storage and sustainable mobility: the study of an isolated system, MIT Portugal Program, Faculty of Engineering, University of Porto
Mott MacDonald (2010) UK electricity generation costs update
Newbery DM (1999) Privatization, restructuring, and regulation of network utilities. The MIT Press, Cambridge
NREL (2011) Renewable energy cost modeling: a toolkit for establishing cost-based incentives in the United States
OFGEM (2010) Handbook for implementing the RIIO model
OMIE (2018) Day-ahead and intraday electricity market operating rules, Madrid
OMIE (2019) Available: http://www.comel.es/en/home/markets-and-products/electricity-market/our-electricity-markets/intraday-market. Accessed 11 July 2019
Palmintier B, Webster M (2011) Impact of unit commitment constraints on generation planning with renewables. Massachusetts Institute of Technology, Engineering Systems Division
Rebours Y, Kirschen D (2005) What is spinning reserve? The University of Manchester, Manchester
REN—Redes Energéticas Nacionais (2004) Plano de Investimentos da Rede Nacional de Transporte 2006–2011: Padrões de Segurança de Planeamento da RNT
REN—Redes Energéticas Nacionais (2008) Report and accounts 2007, Lisbon
REN—Redes Energéticas Nacionais (2009) Report and accounts 2008, Lisbon
REN—Redes Energéticas Nacionais (2010) Report and accounts 2009, Lisbon
REN—Redes Energéticas Nacionais (2011) Report and accounts 2010, Lisbon
REN—Redes Energéticas Nacionais (2012) Report and accounts 2011, Lisbon
REN—Redes Energéticas Nacionais (2013a) Report and Accounts 2015, Lisbon
REN—Redes Energéticas Nacionais (2013b) Report and Accounts 2012, Lisbon
REN—Redes Energéticas Nacionais (2014) Report and accounts 2013, Lisbon
REN—Redes Energéicas Nacionais (2015) Report and accounts 2014, Lisbon

REN—Redes Energéticas Nacionais (2017) Report and accounts 2016, Lisbon
REN—Redes Energéticas Nacionais (2018) Report and accounts 2017, Lisbon
REN—Redes Energéticas Nacionais (2019) Available: https://www.ren.pt/. Accessed 25 July 2019
RWE (2008) Facts and figures 2008, Update October 2008
Schenk KF, Rau NS (1979) Application of fourier transform techniques for the assessment of reliability of generating systems, Paper no. A-79-103-3. IEEE Winter Power Meeting
Sen S, Kothari DP (1998) Optimal thermal generating unit commitment: a review. Electr Power Energy Syst 443–451
Stoft S (2002) Power system economics: designing markets for electricity. IEEE Press & Wiley-Interscience, The Institute of Electrical and Electronics Engineers, Inc.
Stremel JP (1981) Sensitivity study of the cumulant method for calculating generation system reliability, pp 771–78. IEEE transactions, PAS-100
Vazquez M (2006) Optimizing the spinning reserve requirements. PhD Thesis, Faculty of Engineering and Physical Sciences, University of Manchester, Manchester
Viana A (2004) Metaheuristics for the unit commitment problem. PhD thesis, Faculty of Engineering, University of Porto
Wood AJ, Wollenberg BF (1996) Power generation operation and control. Wiley, Hoboken
Zapfel P (2011) Preparing for large-scale carbon allowance auctions. European Commission—Climate Action

Chapter 4
Asset Management Transformation

4.1 Mission and Objectives

4.1.1 Mission and Values

An asset can be deemed in different ways, from a pure financial viewpoint to a technical perspective. In truth, there is a wide list of asset types, such as: physical assets; financial assets; information assets; human assets; etc. Furthermore, an asset is conceived to provide value to an organization (ISO 2014).

According to ISO 55000 (2014), asset management is defined as "coordinated activity of an organization to realize value from assets". Although the assertiveness of this definition, Hastings (2014) introduced a rather detailed description: "given a business or organizational objective, Asset Management is the set of activities associated with identifying what assets are needed, identifying funding requirements, acquiring assets, providing logistic and maintenance support for assets, disposing and renewing assets, so as to effectively and efficiently meet the desired objective".

The asset management mission could vary from sector to sector, and even amongst energy utilities. Yet one could agree on a common mission of asset management for power and gas companies:

> The asset management aims at ensuring the utility's infrastructure reliability and service quality—managing the value at risk via tradeoff analysis—and the assets' integrity, availability and performance, through life-cycle monitoring and TOTEX assessment.

This broad asset management mission for power and gas utilities can be decomposed into three core values, i.e.: (i) sustainable energy systems; (ii) value for money; and (iii) life-cycle assessment (Fig. 4.1).

There is a list of principles that govern the asset management role (ISO 2014):

- Sustainable operation: build a set of state of art processes in respect to safety, quality, health and hygiene, bearing in mind sustainable development concerns;

M. Moreira da Silva, *Power and Gas Asset Management*, Lecture Notes in Energy 72, https://doi.org/10.1007/978-3-030-36200-3_4

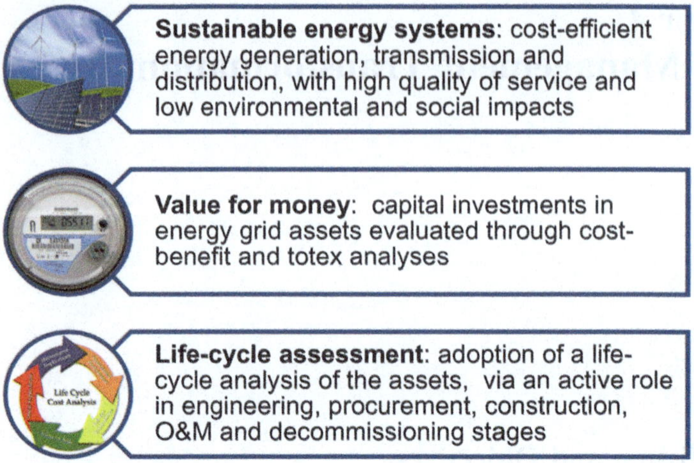

Fig. 4.1 Asset management values for energy utilities

- Continual improvement: strong compromise with continual improvement, establishing processes for asset monitoring and key performance indicators for asset management;
- Value: assets ought to deliver value to the company and its stakeholders;
- Strategic alignment: the asset management should embody the corporate strategy and objectives;
- Leadership and workforce: leadership and organization are key to value creation, by engaging the human resources in the company's strategy (not only in the projects kick-off but via a permanent feedback) and clearly defining roles, responsibilities and empowering asset managers;
- Reliability: the asset management must ensure that assets will fulfill their purpose, by developing and implementing processes that match the corporate objectives with assets' KPIs; ensure the assets' life-cycle integrity; promote continual improvement; and provide the necessary human and financial resources for achieving the asset management targets;
- Risk management: the asset management must be aligned with the risk profile of the company;
- Adequacy: the asset management mission must be consistent with corporate strategy and should be frequently updated.

4.1.2 Strategy

As stated earlier in this chapter, life-cycle assessment is at the core of asset management mission. The assets' end-of-life management involves performing multi-scenario and multi-alternative analyses. An asset manager, definitely, has to deal with uncertainty. Uncertainty on future RoR (for CAPEX), revenue cap (for OPEX), asset performance and technology developments. One way to cope with uncertainty is through scenario analysis. Moreover, the asset manager should envisage a comprehensive portfolio of alternatives for asset end-of-life management, namely:

- asset replacement (CAPEX);
- refurbishment (CAPEX or OPEX, depending of the regulation context);
- or preventive (or corrective) maintenance (OPEX).

The more information, expertise and tools the asset manager holds, the more robust will be the strategy. In view of that, utilities should endeavor a bold corporate transformation, from capital-driven companies to the asset management paradigm. This asset management transformation should encompass the next main levers:

1. Definition of the target operating model;
2. Internal reorganization;
3. Digitalization and data-centric processes;
4. Development of decision models and advanced analytics for predictive asset management;
5. Designing a multi-year asset transformation roadmap to engage the main stakeholders (i.e. shareholders, government and regulator).

4.1.3 Objectives

The objectives of asset management come out coherently with the mission and strategy presented previously. Indeed, an energy utility that starts the asset management transformation journey could list a set of objectives, o, which are measured through attributes, a. This exercise is provided next, for the particular case of an electricity TSO.

o_1: Reliability and quality of service:

$a_{1,1}$: average interruption time;
$a_{1,2}$: energy not supplied;
$a_{1,3}$: system average interruption frequency index;
$a_{1,4}$: momentary average interruption frequency index;
$a_{1,5}$: system average interruption duration index;
$a_{1,6}$: system average restauration index;
$a_{1,7}$: grid flexibility.

o_2: Asset integrity and availability:

$a_{2,1}$: line outages per 1000 circuit km;
$a_{2,2}$: substation outages per 1000 circuit end;
$a_{2,3}$: asset failure rates (minor and major);

$a_{2,4}$: average availability rate of lines and substations;
$a_{2,5}$: average age per asset class;
$a_{2,6}$: average risk index per asset class.

o_3: Efficiency and productivity:

$a_{3,1}$: energy losses;
$a_{3,2}$: line totex per circuit km;
$a_{3,3}$: substation totex per circuit end;
$a_{3,4}$: maintenance orders performed per maintenance orders planned;
$a_{3,5}$: preventive condition-based maintenance orders[1] per total maintenance orders[2];
$a_{3,6}$: investment efficiency[3];
$a_{3,7}$: operation efficiency[4];
$a_{3,8}$: employee turnover ratio;
$a_{3,9}$: share of outsourcing per asset class.

o_4: Energy sustainability:

$a_{4,1}$: RES share in the generation mix (standing as a proxy for the decarbonization);
$a_{4,2}$: SF_6 emissions;
$a_{4,3}$: ratio between the interconnection market capacity and line thermal capacity;
$a_{4,4}$: right of way for overhead lines[5];
$a_{4,5}$: creation of external employment (direct and indirect);
$a_{4,6}$: people and goods safety;
$a_{4,7}$: wholesale energy price variation due to grid investments.

The calculation of these attributes allows building a *tableau de bord* for controlling the accomplishment degree of each objective.

4.2 Operating Model

4.2.1 Overview on Maintenance and Replacement Policies

Overview

Maintenance policies could be distinguished by their timing for action. That is to say, an asset manager should opt for acting before (proactive) or after (reactive) a given equipment failure. The reactive approach is usually referred as corrective

[1]Asset inspections are not considered condition-based maintenance actions.

[2]These include corrective maintenance, but also preventive time-based maintenance and asset inspections.

[3]CAPEX transferred to operation/CAPEX efficiency target.

[4]Actual OPEX/OPEX Revenue Cap.

[5]Metric for the social and environment impact (in km or km^2) of overhead lines.

maintenance. Alternatively, the proactive maintenance seeks to prevent the asset failure and can be implemented by a systematic (time-based) or predictive approach. The predictive maintenance has gained momentum in the last years in the utilities' ecosystem, due to new regulatory models (incentive-based) and privatization processes (leading to an efficiency focus). Although predictive maintenance could be adopted by a wide set of strategies, there are three leading strategies: condition-based maintenance (CBM); risk-based maintenance (RBM); and reliability-centered maintenance (RCM).

Corrective Maintenance

This maintenance strategy addresses the asset just after its failure and aims at recovering the equipment integrity, so that it can be put back into operation. This maintenance approach is often named "run-to-failure" policy. As a matter of fact, corrective maintenance can be a rational strategy even in a predictive maintenance regime. Assets with low criticality could be subject to the run-to-failure approach, following the value-at-risk logic. This strategy could also be reasonable when it is hard to perform condition or risk-based maintenance (owing to lack of historical data or expertise).

Besides these cases, corrective maintenance typically leads to higher costs and energy not supplied, which requires excessive spare parts and backup equipment.

Preventive Maintenance

The principle of preventive maintenance lies on carrying over intermediate maintenance actions, in order to delay the asset degradation and/or preventing the equipment's failure. As referred previously, this maintenance strategy can be clustered into the next types: time-based maintenance; condition-based maintenance; reliability-centered maintenance; and risk-based maintenance.

(i) Time-based maintenance

Time-based maintenance (TBM) is put in place periodically, through planned maintenance actions based on time or operations. This policy is usually set by either the asset manufacturer or the company's O&M division (based on internal expertise). Taking into account that the asset condition is not considered in this maintenance policy, there is the risk of incurring in unnecessary conservation and replacement actions, besides the probability of failure related to the intervention. On the other way around, a given maintenance action could be planned for a moment subsequent to the actual asset failure. Indeed, TBM will likely fail the optimum action point, leading to excessive periodic conservation works and undesired corrective maintenance. Still, in some cases TBM might be the most reasonable strategy, when there is lack of asset condition and performance data (historical and online) or for very complex asset systems. In this case, the equipment manufacturer could provide meaningful information.

(ii) Condition-based maintenance

In the CBM, the asset manager designs a strategy in which the asset is only subject to conservation, refurbishment or replacement actions, if its condition (present or future) puts at risk the grid's reliability, as well as the people and goods safety. The CBM requires the development of deterministic or probabilistic models (based on historical data) to support the annual and multi-annual asset maintenance and replacement plan. For the case of predictive real-time operation, continuous data streams should be monitored at the local SCADA. Given this condition monitoring architecture, the asset manager should build thresholds for condition variables that trigger alarms to be sent to the control room.

CBM seeks to extend the asset end of life, by controlling its life-cycle health and TOTEX. Through CBM, the asset manager could expect higher asset availability, better performance and lower overall costs, since this strategy avoids unnecessary actions and resources. Yet CBM demands the investment in information and communication technologies (e.g. sensors and data collection software) and development of data-centric processes, with comprehensive performance and costs attributes. Besides the advantages of CBM, for both power and gas utilities, it is hard to design a full-condition-based strategy, since there is data and expertise asymmetry amongst the asset base, and there are asset classes that gather a myriad of components difficult to monitor.

Finally, one ought to point out that condition-based maintenance does not include inspection actions. Some maintenance engineers tend to advocate that inspection activities consist of offline condition monitoring. Although historical condition data is key to probabilistic models, actually asset inspections are developed periodically, following a time-based approach.

(iii) Reliability-centered maintenance

RCM is a systematic method for building a maintenance plan, rooted on a detailed process. This maintenance strategy can provide a blended approach (integrating corrective, time-based and condition-based maintenance) and often applies the Failure Mode, Effects and Criticality Analysis (FMECA) (US Department of Defense 2013). According to the International Atomic Energy Agency (2007), RCM potentially decreases periodic maintenance by leveraging CBM. RCM embodies a logical connection between the asset function, the need for reliability and safety, and the concerned maintenance policy, developed through the next guidelines (Hastings 2014):

1. Asset selection and expert team set up;
2. Identification of asset functions, potential faults, failures and degraded states;
3. Identification of failure consequences;
4. Maintenance tasks specification and responsibilities;
5. Maintenance planning.

RCM is particularly helpful for industry-like installations, like refineries and gas stations. Although there are electrical utilities aiming at developing FMECA for substation assets (e.g. TenneT), the vast number and differentiation of power asset classes (and equipment models within each class) turn out RCM tough to implement.

(iv) Risk-based maintenance

RBM is a step further from CBM, standing as a variant approach of RCM. RBM supports the design of maintenance policies and plans, through a model that combines the asset condition (or probability of failure) and its criticality. If the CBM is a one-dimension approach (i.e. asset probability of failure), the RBM draws a two-dimension model (i.e. asset probability of failure and criticality), as sketched in Fig. 4.2.

The decision model of RBM can be mathematically described by a risk index, RI_a^y, presented in Eq. (4.1)

$$RI_a^y = PoF_a^y \times CI_a^y \tag{4.1}$$

where, PoF_a^y consists of the probability of failure of asset a in the year y, whereas CI_a^y stands for the criticality index. The mathematical formulation of both the probability of failure and criticality will be thoroughly discussed in Chap. 5.

Here, risk is utterly addressed as a single attribute, yet it encloses several risk types, i.e.:

- Financial risk;
- Environment risk;
- Safety risk;
- Reputation risk;
- Regulatory risk;
- Energy supply risk.

The risk can be computed through a proxy index, in a continuous or discrete scale. If PoF_a^y and CI_a^y are normalized to a 1 out of 3 scale, then RI_a^y will vary from 1 to 9. Alternatively, risk could be calculated using the attribute units, such as the expected energy not supplied (in MWh). This method allows the risk monetization. In the case of the expected energy not supplied, risk is monetized using the value of lost load (in €/MWh).

According to the RBM model presented in Fig. 4.2, the asset manager is capable to populate each colored quadrant with the concerned assets. Each quadrant embodies a specific maintenance strategy, namely:

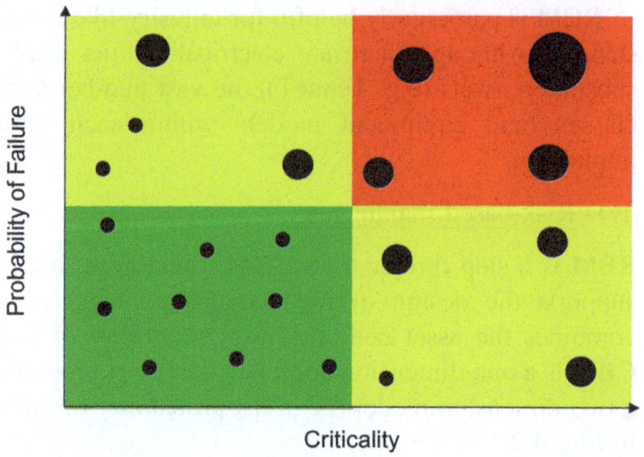

Fig. 4.2 Decision model of RBM

- Red quadrant: urgent maintenance action or asset replacement;
- Yellow quadrant: online condition monitoring (or rather frequent offline inspection, if real-time monitoring is not feasible), and preventive maintenance according to annual plan;
- Green quadrant: lower frequency of asset inspection and maintenance, considering also the "run-to-failure" approach.

Replacement and Refurbishment

After presenting the periodic and predictive maintenance strategies, there is a last category to discuss. The refurbishment is the most invasive maintenance procedure, aiming at bringing the asset's condition close to its original state. It is, hence, an expensive action typically applied once-in-a-lifetime. For instance, a power transformer is a robust and lifelong asset (lasting by 60 years). Still this type of asset is usually subject to a refurbishment action after 30 years in operation. Besides the performance and safety concerns, an asset manager must address the TOTEX, taking into account the regulatory model in place. In this context, the asset refurbishment competes with the replacement option.

The asset replacement could anyway arise as the only management option, when the utility faces at least one of the next motives:

- Refurbishment is no longer an alternative, due to the asset condition;
- Technology obsolescence;
- Lack of internal and external (manufacturers and service providers) know-how;
- High O&M costs.

As addressed in Chap. 3, the regulatory model is a game changer for asset management decisions. In Fig. 4.3 a decision-making logic is proposed for asset maintenance and replacement planning.

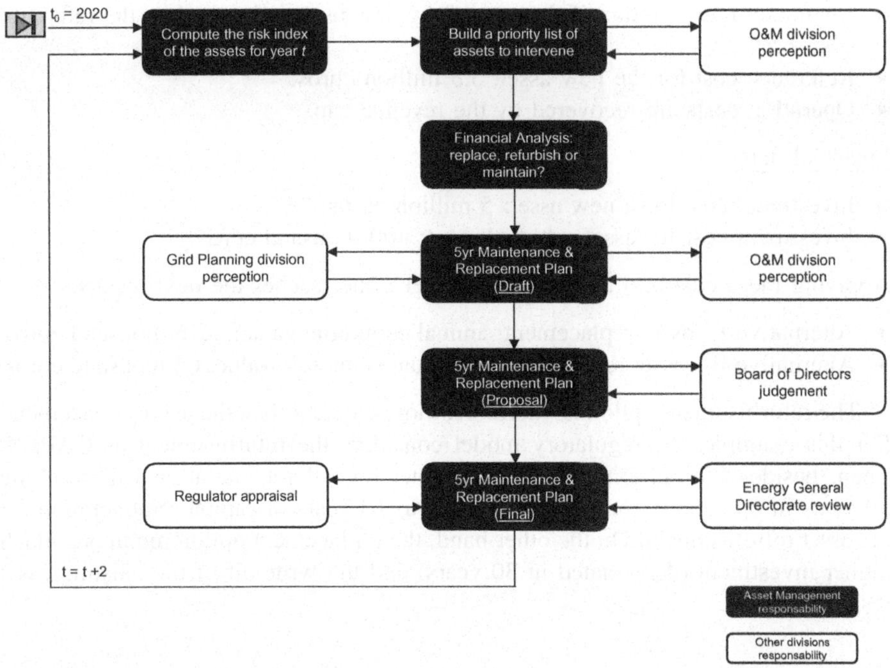

Fig. 4.3 Decision-making logic for asset maintenance and replacement

From the sketched decision-making logic, the asset manager should draw an integrated multi-year maintenance and replacement plan. The risk index supports the development of a priority list of assets to intervene. Yet the intervention encloses different alternatives, i.e.:

(a) Replacement;
(b) Refurbishment;
(c) Maintenance.

Assuming even risk among the alternatives, the financial analysis becomes decisive. To illustrate this analysis, a theoretical example is provided next.

Technical data:

- Asset class: power transformer;
- Transformation power: 450 MVA;
- Voltage levels: 400/220 kV;
- Age: 27 years.

Regulatory data:

- Economic lifetime of a new asset: 30 years;
- Economic lifetime of a refurbished asset: 10 years;
- Regulated rate of return: 6%;

- Regulated rate of return if the cost of the new asset is lower than the reference cost: 6.8%;
- Reference cost for the new asset: 5.5 million euros;
- Operation costs are recovered by the revenue cap.

Financial data:

- Investment cost for a new asset: 5 million euros;
- Investment cost for asset refurbishment: 400 thousand euros.

Applying these data into the financial model[6], one reaches the next results.

- Alternative I—asset replacement: annual economic value: −3.6 thousand euros.
- Alternative II—asset refurbishment: annual economic value: 64 thousand euros.

The refurbishment option is, therefore, more efficient than the asset replacement. For this example, the regulatory model considers the refurbishment as CAPEX. Then, besides the current power transformer annual remuneration (till reaching 30 years in operation), the utility is subject to 10 years of capital remuneration for the asset refurbishment. On the other hand, the replacement option requires a much higher investment (depreciated in 30 years) and the write-off of the existing asset residual value.

4.2.2 Target Model

Having depicted the possible strategies for asset maintenance, it is now the moment to search for the most adequate operating model. Considering the pros and cons of each maintenance policy and the mission and objectives for asset management proposed in Sect. 4.1, the risk-based maintenance arises as the most efficient and sustainable asset management strategy (Figs. 4.4 and 4.5).

The risk-based asset management urges not only a technological investment, but a bold cultural and organization transformation. After drawing the long-run strategy, each utility's board of directors (BoD) ought to design the most suitable operating model. In a simplified way, the BoD face four potential operating models for asset management.

(i) "Grid Expansion" model

This operating model is usually applied by utilities that are in a stage of intense grid growth. In this case, construction and O&M divisions manage CAPEX and OPEX, respectively, in a self-determining manner. Maintenance and replacement activities are casuistically decided by O&M teams, distant to grid planning and regulation divisions, and without an integrated and multi-year plan. Decisions are dominated

[6]The annual economic value is used (instead of net present value) since the two investment alternatives have different time horizons.

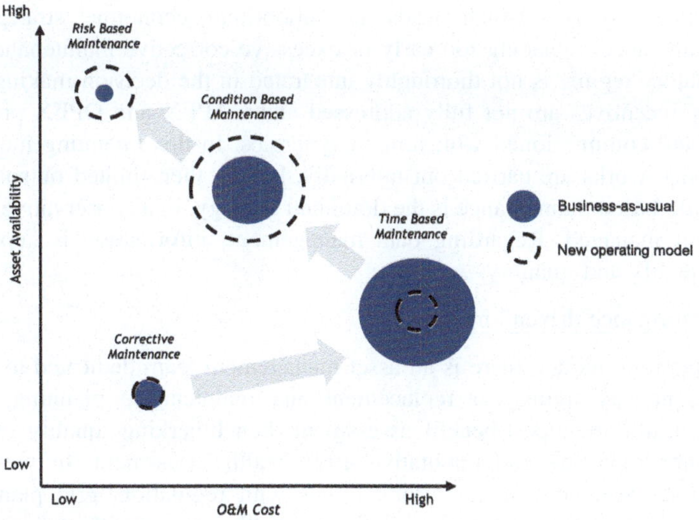

Fig. 4.4 Roadmap for risk-based maintenance (1/2)

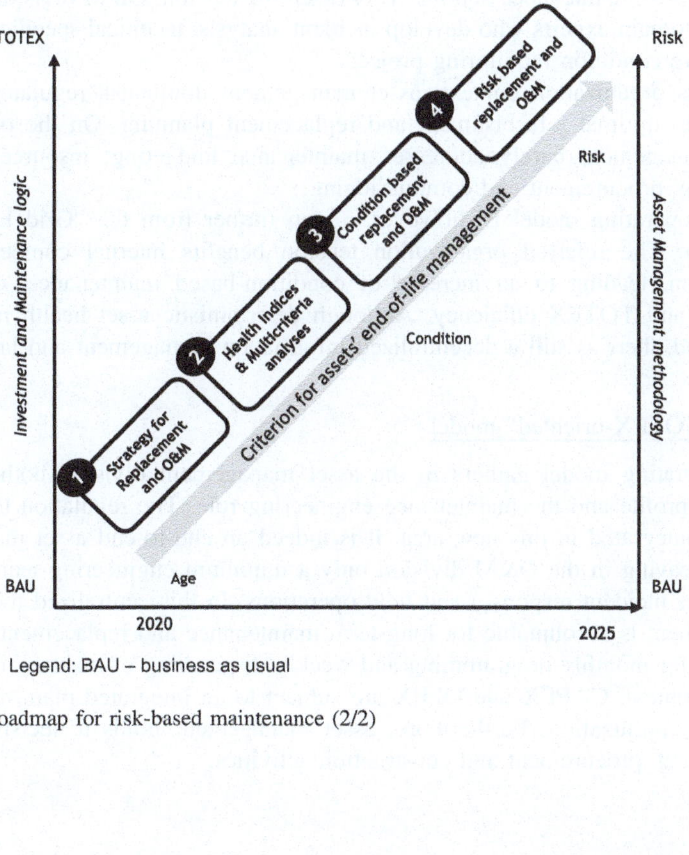

Fig. 4.5 Roadmap for risk-based maintenance (2/2)

by technical criteria, which leads to suboptimal economic strategies (e.g. over-maintenance, replacing too early or excessive corrective maintenance). Since the regulatory regime is not thoroughly integrated in the decision-making process, efficiency incentives are not fully addressed and CAPEX and OPEX projects are planned and commissioned with limited synergies. In this operating model, most maintenance works are carried out internally, having then limited outsourced services. Time-based maintenance is the dominant strategy, in a conservative way, and risk is not managed. Regarding data management, information is kept in silos lacking quality and quantity.

(ii) "Performance-driven" model

In this operating model, there is an asset management team dedicated to upstream activities, namely multi-year replacement and maintenance planning, financial analysis, multicriteria/cost-benefit assessment, benchmarking, quality of service, performance reporting and qualitative asset health assessment. In view of this profile, asset management has strong bonds with regulation, grid planning and design teams. Yet this operating model traditionally raises conflicts between asset management and O&M division, since asset engineering is imbedded in the latter. Indeed, O&M division is responsible for annual planning, budgeting and programming of maintenance actions. To worsen the tension, O&M division typically gathers domain experts who develop incident analysis, technical specifications and innovative condition monitoring projects.

In this organization context, asset management dominates regulatory details, leading to optimal refurbishment and replacement planning. On the other hand, asset management barely addresses maintenance budgeting, insource/outsource balancing, procurement and commissioning.

This operating model is, however, a step further from the "Grid Expansion" approach. The referred organization tension benefits internal competition and innovation, leading to an increase of condition-based maintenance, outsourced services and TOTEX efficiency. Although deterministic asset health models are developed, there is still a decentralized information management and lack of data integrity.

(iii) "TOTEX-oriented" model

This operating model gathers in the asset management division both the asset strategy profile and the maintenance engineering role. The regulation team could also be integrated in this new area. It is indeed an end-to-end asset management model, leaving in the O&M division only a minimum engineering team (e.g. for first-order incident response) and field operations. In this centralized model, asset management is accountable for long-term maintenance and replacement planning, but also for monthly programming and weekly dispatching of maintenance works. In this context, CAPEX and OPEX are subject to an integrated plan, resulting in TOTEX optimization. Furthermore, asset management holds a decisive role in engineering, procurement and construction activities.

As expected, time-based maintenance gives room to condition and risk-based maintenance, depending of the degree of digital transformation and artificial intelligence models. In addition, outsourcing of non-critical activities will likely boost.

This operating model benefits of having—entrenched in the asset management division—a data and analytics team, focused in mathematical modelling (for risk prediction), data engineering, asset sensing and deployment of enterprise asset management software and geographic information system.

(iv) "Portfolio Management" model

Finally, the state of art model for asset management integrates the asset management division (described in the previous point) with grid planning team. Asset management becomes, in this operating model, the key unit of the utility by addressing not only maintenance and replacement decisions, but also grid expansion and reinforcement. There is a caveat though in this model. If the utility is still in a stage of intense grid expansion, the previous operating model is rather adequate (asset management and grid planning separated). On the contrary, if the utility has reached a plateau in grid growth investment and the average age of its asset base is increasing, then the integrated asset management and grid planning approach becomes more efficient.

In this framework, there is an obvious synergy between asset data analytics with grid simulation. This blended data science team could leverage not only the stochastic models for risk-based maintenance, but also move from the deterministic "n − 1" criterion to probabilistic grid planning.

For multi-utilities (managing both power and gas grids), this operating model is characterized by integrating in the asset management division power and gas processes, establishing synergies in the fields of maintenance policies, budgeting, cost-benefit analyses, benchmarking, reporting and data analytics.

Based on this overview and on the risk-based maintenance principles, the target operating model for asset management could be both the "TOTEX-oriented" or the "Portfolio Management". To simplify, let's assume the more ambitious one, i.e. the latter.

The development of the "Portfolio Management" model implies the next transformation actions.

1. Internal reorganization of teams and processes:
   - merge of power and gas business units (when applicable);
   - integrated division for grid planning and asset management.
2. Design of coherent decision-making models (aligned with the regulatory regime):
   - policy and strategy for asset maintenance and replacement;
   - strategy for spare parts management;
   - insource/outsource strategy per asset class;
   - risk model development for investment prioritization;

- multicriteria/cost-benefit methodology for grid growth and replacement projects.

3. Asset digitalization and advanced data analytics:

   - roadmap for asset sensing;
   - data centralization and governance;
   - definition of data models and dictionaries;
   - development of predictive maintenance algorithms;
   - deployment of a probabilistic grid planning model;
   - development of a model for optimized replacement planning.

4.3 Governance and Organization

The asset management transformation—towards the "Portfolio Management" model—requires the organizational and cultural change, as referred in Sect. 4.2. Firstly, the utility's BoD should clearly distingue the next functions:

- Asset owner: entity responsible for strategic business planning, corporate finance, risk and sustainability;
- Asset manager: entity accountable for implementing the business strategy defined by the asset owner, by designing and controlling the TOTEX plan in order to achieve the desired financial and performance outputs;
- Asset service provider: internal or external entity responsible for the execution of TOTEX plan drawn by the asset manager, guaranteeing the assets' integrity, people and goods safety, and data collection and sharing, using the prescribed time and resources.

Setting up these functions demands reshaping the company's organization, in a top-down approach. One provides the traditional utility's organization for each of the operating models in Figs. 4.6, 4.7, 4.8 and 4.9.

The proposed company reorganization centralizes into the Asset Management area the utility's core decision-making processes, leaving to the Operations area tasks that can be subject to outsourcing (i.e. asset service provider). The Finish TSO (Fingrid) is a good example of centralized asset management, with outsourced grid construction and maintenance, crafting an outstanding performance (technical and financial) (Fingrid 2018). Regarding the system operation department, this could be handled either by the utility (in the TSO and DSO models) or via an external party (e.g. state-owned organization).

Finally, both the Corporate Center and the Business Development areas stand as the asset owner entity, referred in the beginning of this section.

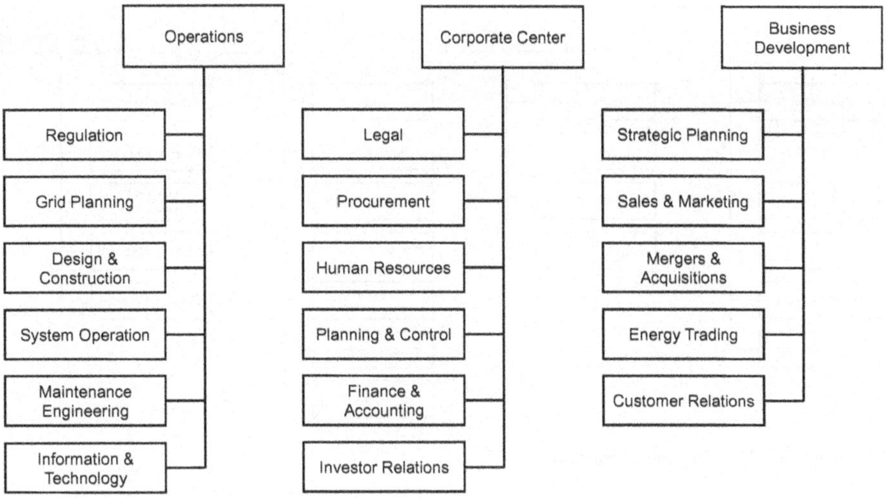

Fig. 4.6 "Grid Expansion" model

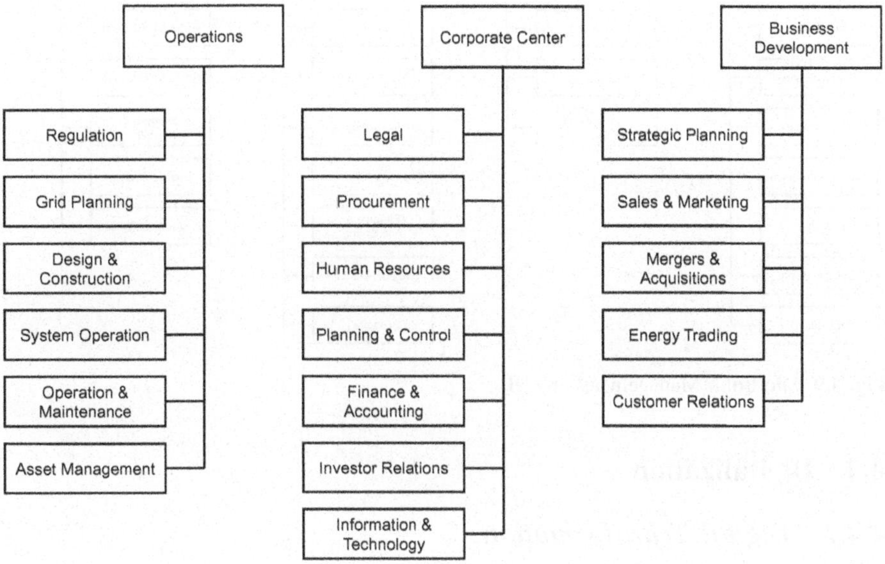

Fig. 4.7 "Performance-driven" model

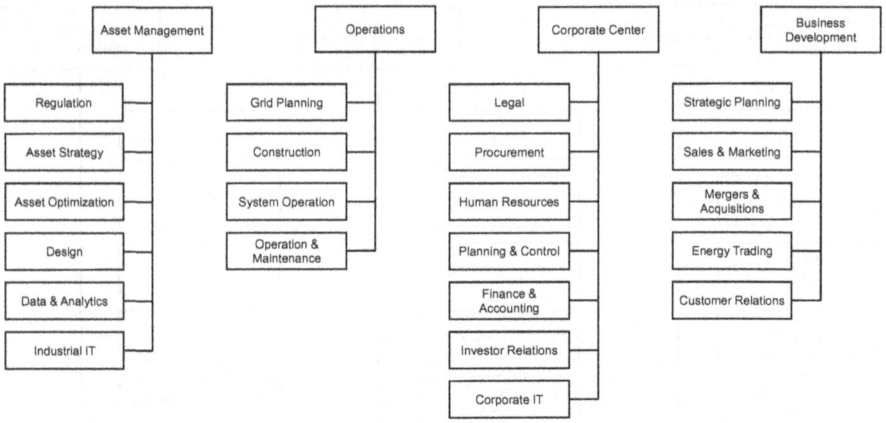

Fig. 4.8 "TOTEX-oriented" model

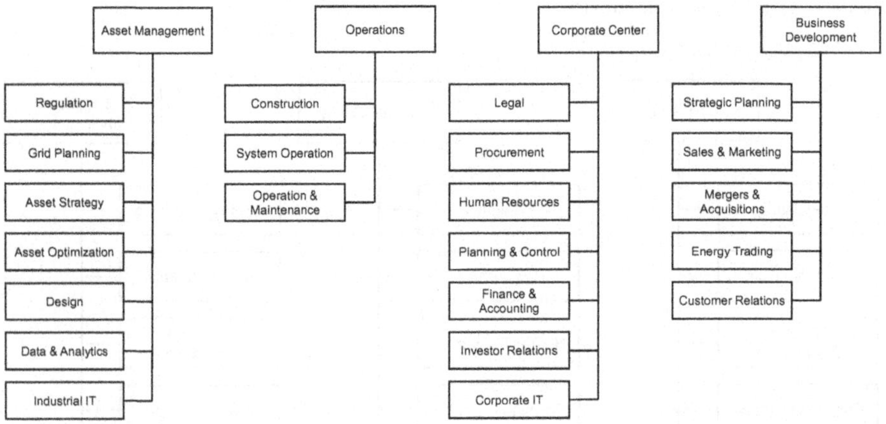

Fig. 4.9 "Portfolio Management" model

4.4 Digitalization

4.4.1 Digital Transformation

Digitalization is a top-ranked topic in world's developed economies. Digital technologies have gained momentum over analog ones, fueled by high-speed data communication, microelectronics and the "internet of things". Still digital transformation arises together with a myriad of innovative technologies and concepts, such as: image processing; visual and audio recognition; blockchain; cybersecurity; quantum computers; robotics, augmented and virtual reality; artificial intelligence; unmanned autonomous vehicles; and 3D printing.

The energy industry is not an except to the "going-digital" trend. In truth, there is a set of technologies and processes, in the energy sector, that depend of the utter implementation of digital systems, namely:

- Advanced meter infrastructure;
- Microgrids;
- Battery energy storage;
- Automated inspection of overhead lines, substations, pipelines and gas stations;
- Predictive vegetation management;
- Asset online condition monitoring;
- Asset and installation redesign;
- Building information modelling;
- Reinforced TSO-DSO coordination;
- Dynamic retail pricing;
- Accurate forecasting of load, variable RES and operating reserve;
- Integrated system, grid and asset operation;
- Real-time digital simulation of energy systems;
- Workforce planning (optimal dispatching);
- People analytics;
- BigData management.

For the specific case of electric utilities, there is a progress imbalance between TSO and DSO (as a result of their purposes), being the former rather advanced in the digital transformation (Fig. 4.10).

Digitalization stands as one of the key levers for asset management transformation. Basically, it seeks to increase the data quality, quantity and integrity, as a means to continuously monitor the assets' condition and build decision models towards risk-based maintenance. In Fig. 4.11 there is a shortlist of the acutest drivers and impacts of digitalization for asset management.

Despite the benefits of asset digitalization, there are no "free lunches". Indeed, the digital transformation ought to be subject to a comprehensive cost-benefit analysis. For instance, for some asset classes there is a threshold for sensing feasibility. Beyond a given number of sensors, there is no investment recovery.

This concern suggests the development of business cases for each asset class, as a preparation to the digitalization roadmap. The business case could offer the decision-maker the next information template:

1. Asset class profile: number of assets; average age; technology types; manufacturers; location;
2. Performance: number, type and causes of incidents and failures; failures per technology and manufacturer;
3. Condition variables: identification of the asset variables for condition monitoring;

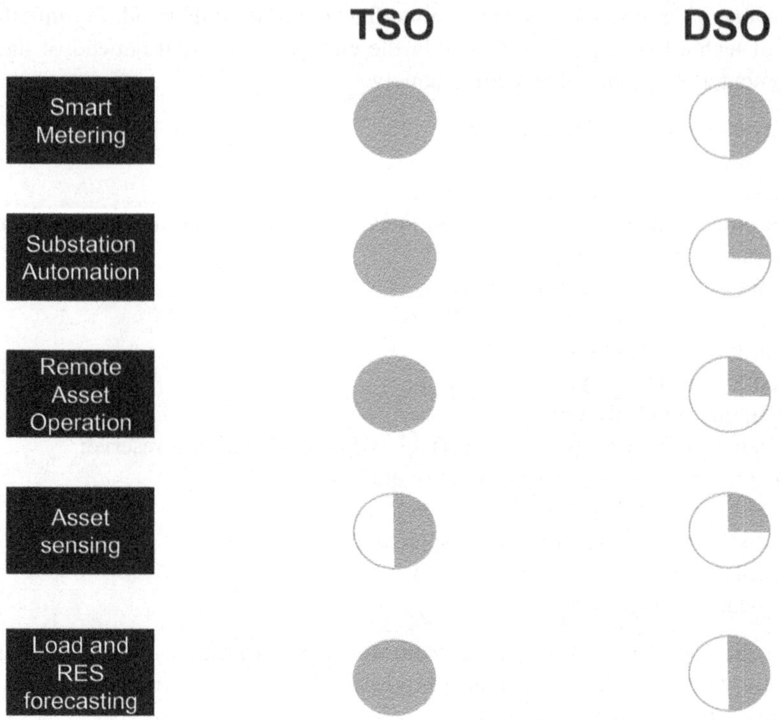

Fig. 4.10 Simplified digitalization progress in electricity TSOs and DSOs

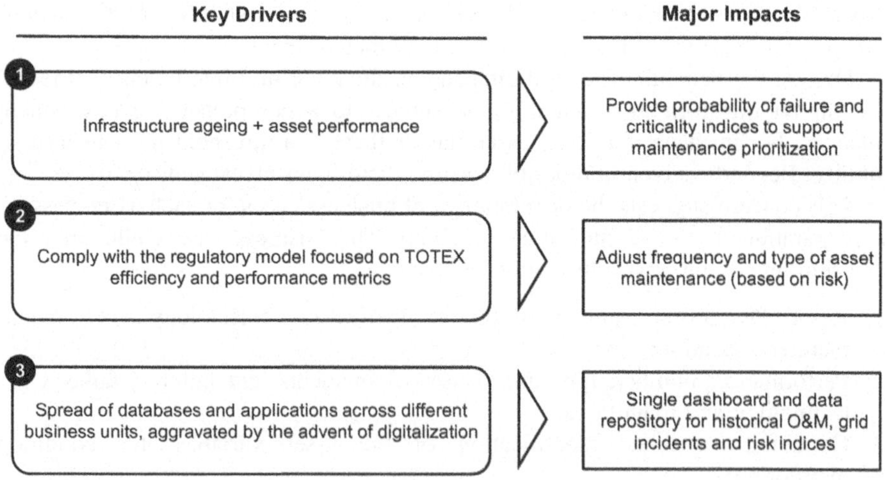

Fig. 4.11 Drivers and impacts of digitalization for asset management

4. Data collection and management: identification of online and offline data sources and warehouses, and condition variables that are not being monitored;
5. TOTEX assessment: financial analysis of the investment and operation for different scenarios of asset sensing (i.e. monitored variables and asset population);
6. Benefit analysis: monetization of the expected benefits driven by asset digitalization;
7. Proposal: cost-benefit analysis coupled by an implementation roadmap.

To illustrate this methodology, a theoretical business case is presented for power transformers' digitalization.

1. Asset class profile

   - number of assets: 200;
   - average age: 25;
   - technology types: 63 MVA; 126 MVA; 170 MVA; 360 MVA; and 450 MVA.

2. Performance

   - Incident causes (and corresponding proportion) (Bartley and William 2003):

     - Insulation (26%);
     - Design/material (23%);
     - Unknown (16%);
     - Loose connection (6%);
     - Overloading (5%);
     - Improper O&M (5%);
     - Oil contamination (4%);
     - Line surge (4%);
     - Lightning (3%);
     - Fire/explosion (3%);
     - Flood (2%);
     - Moisture (1%).

   - Faulty component (and corresponding proportion) (Sokolov 2006):

     - Windings/insulation (29%);
     - Terminals (29%);
     - Overload tap-changer (13%);
     - Tank and dielectric fluid (13%);
     - Magnetic circuit (11%);
     - Bushing (5%).

3. Condition variables
   - Asset register:
     - Age;
     - Technology;
     - Application type;
     - Installation;
     - Manufacturer.
   - Performance:
     - Number of incidents;
     - Number of failures (major and minor).
   - Maintenance:
     - Number and type of maintenance actions;
     - Cost of maintenance actions (per type).
   - Operation:
     - Active and reactive power;
     - Number of maneuvers of the regulator.
   - Electrical measurements:
     - Current at the primary and secondary sides (3 phase) and neutral current;
     - Voltage at the primary and secondary sides (3 phase).
   - Oil temperature:
     - Top oil temperature;
     - Thermal imaging;
     - Environment temperature.
   - Electrical insulation (at bushings and windings):
     - Dielectric dissipation factor;
     - Capacitance (C1);
     - tan δ_{C1};
     - Insulation resistance.
   - Oil quality:
     - Dissolved gases (H_2O, H_2, O_2, N_2; CO, CO_2, CH_4, C_2H_2, C_2H_4, C_2H_6);
     - Degree of polymerization;
     - Furanic compounds.

4. Data collection and management: online monitoring of operation and electrical variables (collected from SCADA); asset register, performance and maintenance data are stored in the enterprise asset management (EAM) system; and the other variables are collected offline during in-field asset inspection and testing.

5 & 6. TOTEX assessment and benefit monetization

- Regulatory model:
  - OPEX for monitoring system is not covered by the revenue cap;
  - OPEX for offline periodic inspections is considered in the revenue cap;
  - Regulated rate of return = WACC = 6%;
  - Depreciation period = 10 years.
- Reliability data [assumptions from (Cigré 2015)]:
  - Power transformer without online monitoring
    - Probability of a catastrophic failure = 0.0007;
    - Probability of a non-catastrophic failure = 0.0063.
  - Power transformer with online monitoring
    - Probability of detecting (earlier) a failure = 0.0042;
    - Probability of a catastrophic failure = 0.00028;
    - Probability of a non-catastrophic failure = 0.00252.
  - Maintenance/refurbishment cost due to a major failure = 200 thousand euros;
  - Maintenance/refurbishment cost due to a catastrophic failure = 1.5 million euros;
  - Predictive maintenance (earlier detection due to online monitoring) = 40 thousand euros.
- Decision-making alternatives
  - Decision hypothesis: sensing investment on 20 power transformers (among the 200 units fleet).
  - Alternative I—"Business-as-usual" monitoring, i.e. no additional sensing:
    - Annual inspection for dissolved gas analysis (DGA) = 1000 €/unit/yr (OPEX);
    - Electrical insulation testing = 5000 €/unit/5 yr (OPEX);
    - **TOTEX = 8.6 million euros.**

- Alternative II—"Low-range" online monitoring, i.e. single-gas DGA (H_2), temperature and bushings:
    - DGA annual inspection = 1000 €/unit/yr (OPEX);
    - DGA online monitoring = 8000 €/unit (CAPEX);
    - Electrical insulation testing = 5000 €/unit/10 yr (OPEX).
    - Bushing online monitoring = 20,000 €/unit (CAPEX);
    - Temperature online monitoring = 0 €/unit (using Pt100 available in the machine);
    - Monitoring system maintenance = 1000 €/unit/yr (OPEX).
    - **TOTEX = 9.0 million euros**;
    - **Annual economic value = 12.2 thousand euros**;
    - **Investment rate of return = 10%**.
- Alternative III—"Mid-range" online monitoring, i.e. DGA with 3 or 5 gases, temperature and bushings:
    - DGA annual inspection = 1000 €/unit/yr (OPEX);
    - DGA online monitoring = 25,000 €/unit (CAPEX); ½ full time equivalent for 20 monitored units (OPEX);
    - Electrical insulation testing = 5000 €/unit/10 yr (OPEX).
    - Bushing online monitoring = 20,000 €/unit (CAPEX);
    - Temperature online monitoring = 0 €/unit (using Pt100 available in the machine);
    - Monitoring system maintenance = 1000 €/unit/yr (OPEX);
    - **TOTEX = 9.6 million euros**;
    - **Annual economic value = −9.9 thousand euros**;
    - **Investment rate of return = 4%**.
- Alternative IV—"High-range" online monitoring, i.e. DGA with 9 gases, temperature and bushings:
    - DGA annual inspection = 0 €/unit/yr (9-gas DGA dismisses this inspection);
    - DGA online monitoring = 50,000 €/unit (CAPEX); 1 full time equivalent for 20 monitored units (OPEX);
    - Electrical insulation testing = 5000 €/unit/10 yr (OPEX).
    - Bushing online monitoring = 20,000 €/unit (CAPEX);
    - Temperature online monitoring = 0 €/unit (using Pt100 available in the machine);
    - Monitoring system maintenance = 1000 €/unit/yr (OPEX);
    - **TOTEX = 10.1 million euros**;
    - **Annual economic value = −13.5 thousand euros**;
    - **Investment rate of return = 4%**.

7. Proposal

- According to the previous multi-alternative analysis, there are a few conclusions to point out:
  - The determining factor in the transformer digitalization relates with the DGA's monitoring range;
  - Alternative II ("low range" online monitoring) dominates the remaining ones, since it leads to lower TOTEX and higher annual economic value and rate of return;
  - A risk analysis (considering the assets' probability of failure and criticality) should be carried out, to select the 20 riskiest transformers to be subject to online monitoring;
  - The asset manager though could opt for a different approach, by clustering these 20 riskiest transformers in 3 segments, each one with a distinct online monitoring alternative.

Following this business case approach, the asset manager is skilled to draw the digital transformation roadmap. The next figures propose a possible asset digitalization scheduling (where *y* refers to year), based on the value and risk of each asset class (Figs. 4.12 and 4.13).

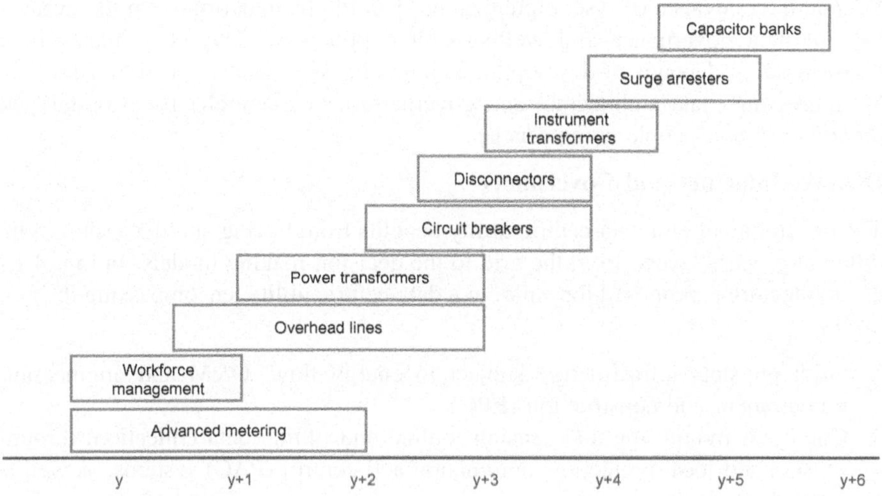

Fig. 4.12 Roadmap for power asset digitalization

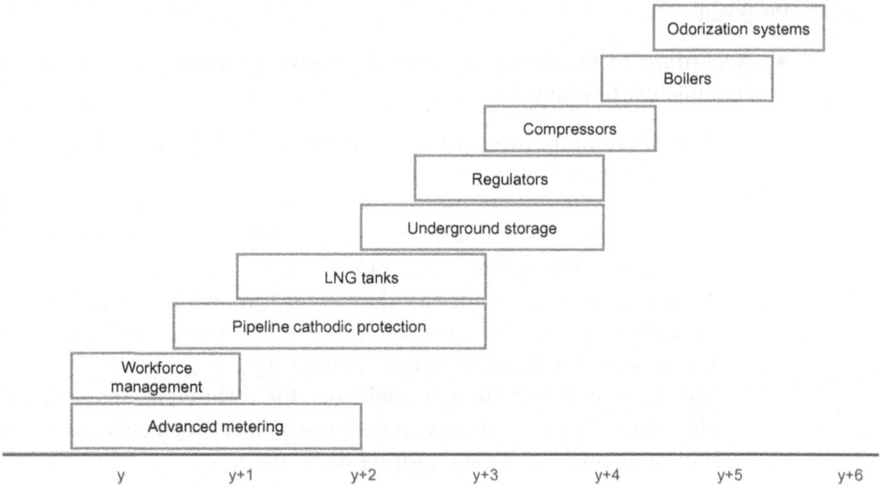

Fig. 4.13 Roadmap for gas asset digitalization

4.4.2 Data-Centric Organization

Despite the relevance of asset digitalization, it could be undermined in the event of lack of data governance and well-structured processes. The asset management transformation depends of developing a comprehensive data and analytics journey, by addressing: data architecture and governance; data centralization, models and dictionaries; and people competencies.

Data Architecture and Governance

The development of a data-centric utility benefits from having an utter vision of the different process layers, from the grid to the decision-making models. In Fig. 4.14, an architecture is proposed for building a data-centric utility, encompassing the next layers:

1. Grid: physical infrastructure subject to energy flow, O&M and engineering, procurement and construction (EPC);
2. Condition monitoring and sensing: online and offline data collection, through sensors and local protection, automation and control (PAC) systems, as well as via in-field asset inspection;
3. Data integration and storage: the enterprise asset management (EAM) system provides static data of the assets (e.g. age, installation, manufacturer, technology type) and maintenance information (e.g. failures, maintenance orders, O&M plans, inspections results), whereas SCADA stores historical records of incidents/outages, events and operations (e.g. asset maneuvers);

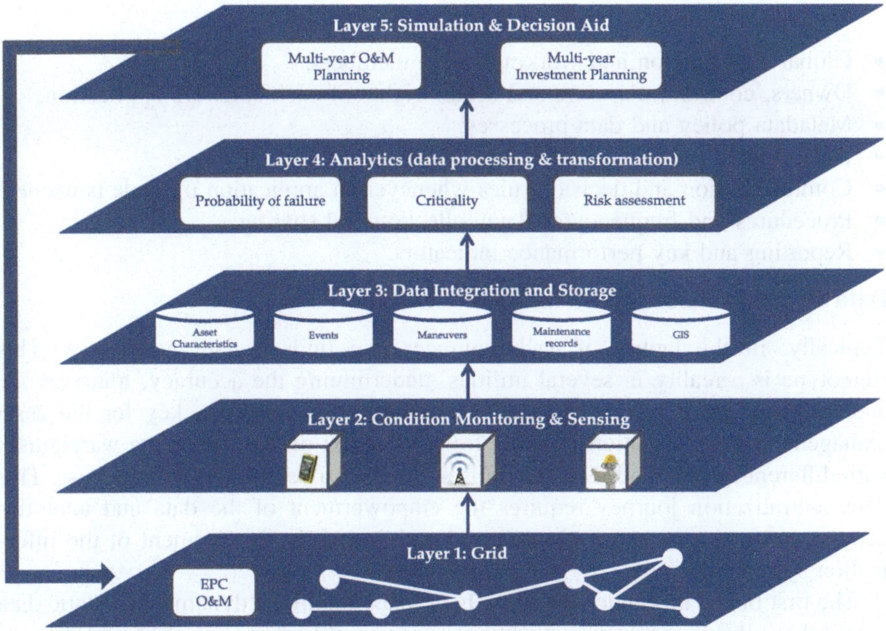

Fig. 4.14 Architecture of a data-centric utility

4. Analytics: this layer is accountable for developing deterministic and probabilistic risk models, focused in the short (less than one year), medium (between 1 and 3 years) and long-term (more than 3 years) analyses;
5. Simulation and decision-aid: investment and maintenance plans based on financial constraints, regulatory model, workforce and risk indices.

Besides the traditional structured data for asset management (failures, incidents, maintenance orders, etc.), the utility should also prepare the data infrastructure for handling unstructured data, such as asset real-time attributes, weather variables, corrosion indices, vegetation growth, lightning, etc. The structured data should be collected through extraction, transformation and loading (ETL) tools, and stored in a data warehouse. Regarding the unstructured data, a data ingestion tool should be used together with a data lake, since flexibility and scalability are key for these types of variables.

Taking into account the wide variety of data sources and internal users, the robustness of the decision aid models requires the adoption and implementation of a holistic data governance. A centralized data and analytics department should be responsible for drawing the mission and rules for the utility's data governance, as well as ensuring periodic meetings for alignment.

This data governance must include the next information:

- Global coordination and working group members;
- Owners, contributors, users and access rights of each software application;
- Metadata policy and data processes;
- Data models and dictionaries (e.g. unique asset ID);
- Communication and decision rules whenever an application upgrade is needed;
- Procedures and frequency of data collection and sharing;
- Reporting and key performance indicators.

Data Centralization, Models and Dictionaries

Typically, in the laptop of each engineer one finds a data warehouse. This stereotype is a reality in several utilities, undermining the accuracy, transparency and accessibility of business data. Data centralization is, then, key for the asset management transformation. It starts from a context of countless data warehouses with different models and IDs, and aims at building a centralized data process. This data centralization journey requires the empowerment of the data and analytics department (acting as a "single point of truth") and the engagement of the information systems division.

The first tier of this endeavor consists of mapping all the dynamic and static data across the utility's software applications, as exemplified in Figs. 4.15 and 4.16.

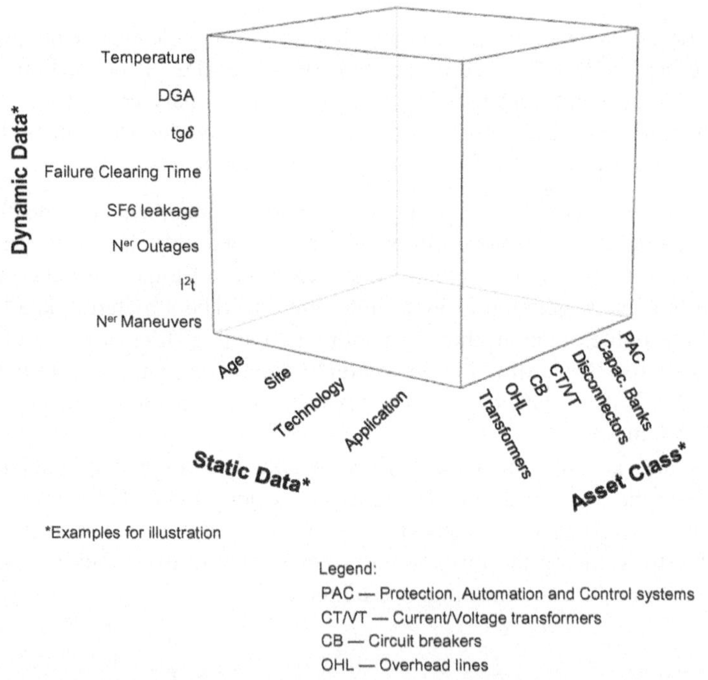

Fig. 4.15 Asset data model

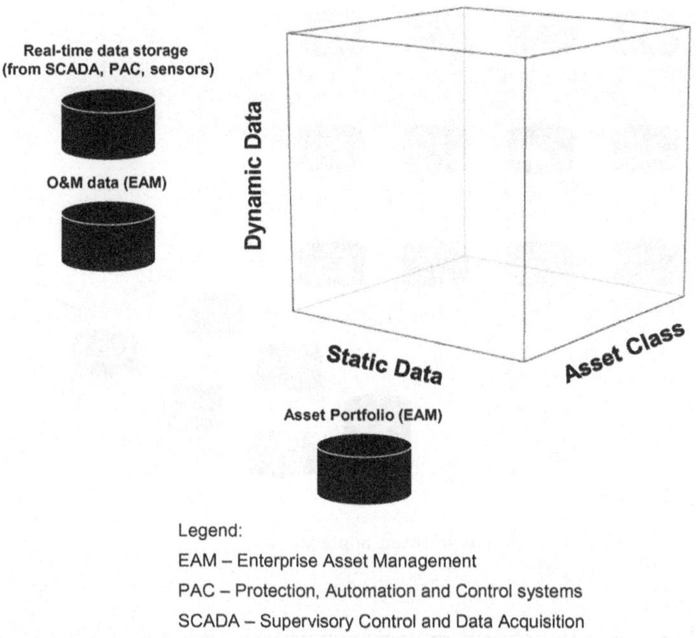

Fig. 4.16 Data mapping

The data centralization involves also identifying the asset ID in each application (illustrated in Fig. 4.17), which will enable building up data dictionaries and analytical models (Fig. 4.18).

Finally, data dictionaries must be written in order to standardize the asset information across the company's applications and ensure data quality processes. The data dictionary should have at least the next contents.

- Asset definition:
  - Label/ID (functional name of the asset);
  - Full name of the asset;
  - Serial number;
  - Asset class;
  - Installation;
  - Manufacturer;
  - Commissioning year;
  - Risk index;
  - Type of data creation (manual or automatic).
- Data governance:
  - Data owner;
  - Data creator;
  - Access rights.

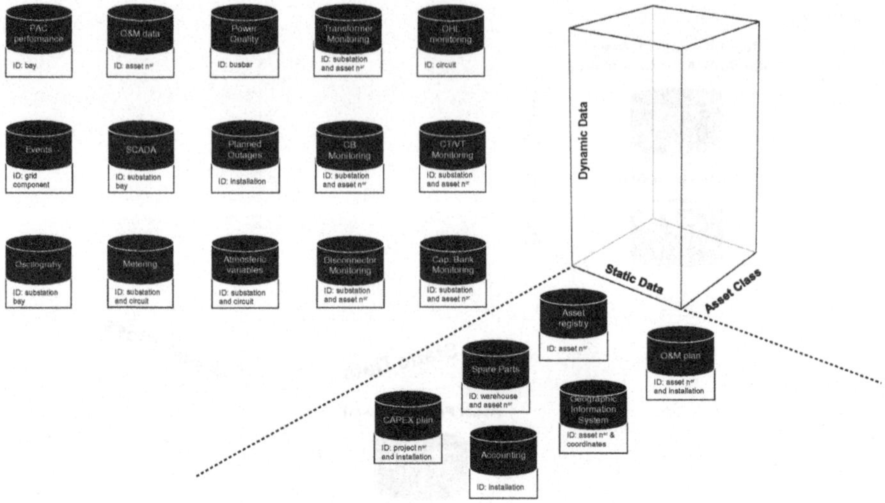

Fig. 4.17 Asset ID in the different software applications

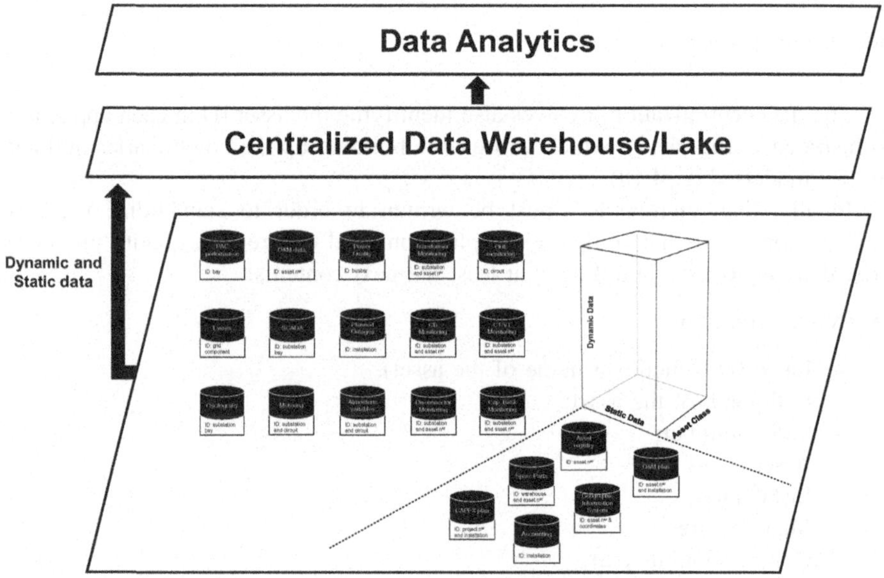

Fig. 4.18 Data centralization

- Data storage and accessibility:
  - Data source (e.g. SCADA);
  - Data type (csv, xls, etc.);
  - File name;
  - Table name;
  - Data field.

People and Competencies

Besides the process and technology transformation towards a data-centric organization, the workforce should be reskilled. In a digital-driven utility, traditional engineering roles—such as design, grid planning and system operation—will turn out commodities, due to the advent of artificial intelligence, robotics and automated processes. Consequently, utilities ought to carry out a strategic workforce plan, by matching the business trend (i.e. asset management transformation) with people's skills.

The utility of the future will be surely sustained by engineers with strong digital competencies, including programming. Designing overhead lines or planning grid expansion will require the use of machine learning and optimization methods.

If the general utility's workforce will have to become rather digital-oriented, a dedicated data and analytics team should be set up, gathering way different profiles. This team should be led by a manager with utility-based experience, though holding a strong analytical background and computer science proficiency. The data and analytics team will benefit of blending domain experts (e.g. former grid planners and maintenance engineers) with data scientists. While the former could be found internally, the latter will most probably be hired externally. Data scientists are both critical for the asset management transformation and hard to find in the market. Indeed, this role compromises an odd combination of skills, i.e.: business know-how; statistics and mathematics; and data engineering and programming.

The set-up of the data and analytics will enable the utility to develop, internally, open-source and tailor-made models, since there is lack of off-the-shelf solutions (from vendors) to comply with the requirements for asset management transformation. Furthermore, internal modelling and programming reduces the risk of depending on external parties to perform core functions, besides enhancing critical mass.

References

Bartley PE, William H (2003) Analysis of transformer failures. In: International association of engineering insurers, 36th annual conference, Stockholm

Cigré (2015) Guide on transformer intelligent condition monitoring (TICM) systems, 630. Working Group A2.44

Fingrid (2018) Fingrid debt investor presentation

Hastings NAJ (2014) Physical asset management with an introduction to ISO55000, 2nd edn. Springer, Berlin
International Atomic Energy Agency (2007) Implementation strategies and tools for condition based maintenance at nuclear power plants. International Atomic Energy Agency
ISO—International Organization for Standardization (2014) ISO 55000, Asset management—Overview, principles and terminology
Sokolov V (2006) Failure statistics. Transformer and bushings design review. Typical failure modes and failure causes. What can be learned from post mortem inspection. In: Fifth AVO New Zealand international technical conference, Methven
US Department of Defense (2013) Defence logistics support chain manual—supportability engineering—maintenance design

Chapter 5
Decision Models and Advanced Analytics

5.1 Introduction to Decision Aid

"Decision aiding is the activity of the person who, through the use of explicit but not necessarily completely formalized models, helps obtain elements of responses to the questions posed by a stakeholder of a decision process. These elements work towards clarifying the decision and usually recommending, or simply favoring, a behavior that will increase the constituency between the evolution of the process and this stakeholder's objectives and value system." This definition was provided by Roy (1996). A decision is related to the comparison of different points of view (some in favor and some against a given decision), which can be approximately defined as criteria. For many years this multicriteria nature of a decision was disregarded by the use of single-criterion methodologies. Nonetheless, for at least thirty years a new look at decision problems has gained importance, by evaluating the pros and cons of several points of view. This methodology is generally named Multicriteria Decision Aid or Multicriteria Decision Analysis (MCDA). The basic principles of MCDA are very clear: a finite or infinite set of actions (alternatives, solutions, courses of action); at least two criteria; and at least one decision-maker (Figueira 2005; Catrinu 2006).

The process of planning power and gas systems encompasses the formulation of a decision. In view of that, the energy system should be modelled and projected in order to be subject to a decision by the decision maker. The problem can be either operation-based or strategy-based, depending on the hierarchical level of the decision. Typically, operation problems seek to optimize the exploitation of a given resource, for performing a pre-defined task. These problems have been previously evaluated by the decision makers, who have defined the tasks, resources and rules of the problem. Alternatively, in strategic planning the decision maker is actively involved and the problems are rather abstract, which requires the identification of objectives, criteria and alternatives. Therefore, identification of problem nature is

M. Moreira da Silva, *Power and Gas Asset Management*, Lecture Notes in Energy 72, https://doi.org/10.1007/978-3-030-36200-3_5

fundamental for an adequate separation between operation optimization and decision aid (Matos 1988).

Decision aid requires that a particular stakeholder is identified as decision maker, who plays a critical role in the process evolution. Nevertheless, the decision maker might be a spokesperson of third parties. The decision aid is usually developed by an analyst, who should be a specialist in charge of building the model (Roy 1996). According to Chankong and Haimes (2008), the MCDA starts with the decision maker's perception of the need for change regarding the system about which he/she is committed. Afterwards, the decision situation will be diagnosed, by identifying the decision variables, alternatives and criteria. At the stage of the problem formulation, the vaguely stated criteria ought to be translated into an operational set of specific multiple criteria. Taking into consideration that the alternatives must be compared, a set of attributes should be defined. These attributes—whose values can be obtained from the model or by subjective judgements—are used as benchmarks of the attainment degree of the criteria.

After the problem formulation step, a well-tuned model should be built-up. A model is understood as a collection of key variables and their relationships, which together can comprehensively tackle the problem under analysis. One of the functions of the model is to generate alternative courses of action. The assessment and evaluation steps are finished when each alternative is evaluated relative to others (Chankong and Haimes 2008).

Roy (1993) defined the next types of multicriteria problems.

- Choice problems: when a simple choice must be made from a set of possible actions (or decision alternatives).
- Sorting problems: when actions must be sorted into classes or categories such as "definitely acceptable", "possibly acceptable but needing more information" and "definitely unacceptable".
- Ranking problems: when actions must be ranked according to some sort of preference order, which might not necessarily be complete.
- Learning (descriptive) problems: when actions and their consequences must be described in a formalized manner so that decision makers can evaluate them. These are essentially learning problems, in which the decision maker aims to get a larger understanding of what may or may not be achievable.

Belton and Stewart (2002) have added two more types of multicriteria problems, as follows.

- Design problems: these problems imply searching, identifying or creating new decision alternatives to meet the goals and aspirations identified through the MCDA process. Keeney (1992) endorsed this methodology, through the concept of "value focused thinking".
- Portfolio problems: when a subset of alternatives must be chosen from a large set of possibilities, taking into consideration both the characteristics of the individual alternatives and the manner in which they interact, as well as the positive or negative synergies between them.

In terms of taxonomy, this book follows the formulation presented in Hwang and Masud (1979), Clímaco (1981) and Matos (1988), when it comes to the classification of multi-objective/multi-attribute. These authors stated that decision aid problems can be classified into two categories:

- Multi-attribute (MA) problems;
- Multi-objective (MO) problems.

The MA problems are usually enclosed to a few predetermined alternatives (e.g. maintain, replace or refurbish a given asset). The alternatives embody an attainment level of the attributes, based on which the final decision is made. The final selection of the alternative is made with the help of inter and intra-attribute comparisons. On the other hand, MO problems aim at computing the "best" alternative, by considering the various interactions within the design constraints which best satisfy the decision maker, by way of attaining some acceptable levels of a set of quantifiable objectives (Hwang and Masud 1979).

The risk model proposed in this chapter entails the next framework.

- Decision maker: chief asset manager;
- Stakeholder(s): asset expert(s);
- Outcome: asset maintenance and replacement plan, through risk prioritization;
- Time horizon: current and forthcoming years;
- Number of alternatives: finite (e.g. replace, maintain or refurbish);
- Type of problem: multiattribute.

5.2 Risk Model

The ultimate layer in the data architecture for asset management transformation (presented in Fig. 4.14) refers to decision aid models. These models materialize valuable information road to risk-based asset maintenance and investment. Within the utility's business, the decision aid models vary according to the purpose and time-horizon, as illustrated in Fig. 5.1.

The asset manager grasps a specific time-bandwidth, from weeks to decades. The asset management activities should rely on a risk model, to allow maintenance and replacement prioritization. The proposed risk model is based on Fig. 4.2 and Eq. (4.1), presented in Sect. 4.2.1, where RI_a^y stands for the risk index of asset a in the year y.

The risk index results from the combination of probability of failure, PoF_a^y, with criticality index, CI_a^y. Although probability of failure is the most used methodology for risk computation, this book proposes an alternative approach, specifically the fragility index, FI_a^y.

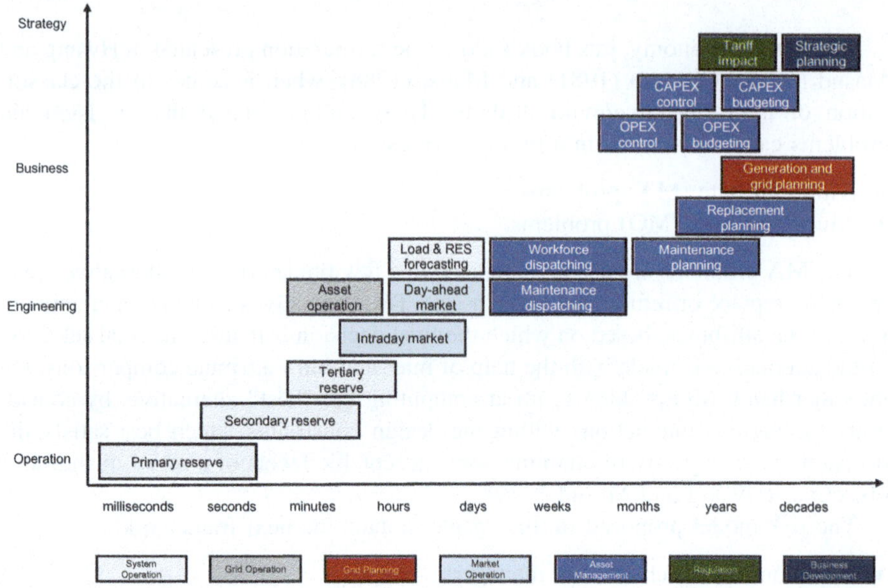

Fig. 5.1 Purpose and time-horizon of decision aid models

Taking into consideration the lack of data-centricity in most utilities and asset complexity in the energy industry (rather systems of assets), probability density functions should be used farsightedly. In truth, probability distributions in these circumstances embody a meaningful inaccuracy. Accordingly, one hereby proposes a fragility index somehow inspired in the work of Taleb (2012).

According to Taleb (2012), "fragility is related to how a system suffers from the variability of its environment beyond a certain preset threshold". Furthermore, this author advocates that "we can gauge the response of a given object to the volatility of an external stressor that affects it".

The fragility index, FI_a^y, proposed in Eq. (5.1), is a proxy metric for the asset probability of failure.

$$\forall_a \, FI_a^y \in FI \subset \mathbb{R} \wedge L \leq FI \leq U \tag{5.1}$$

where L stands for the lower limit of the index, whereas U refers to the upper limit.

From this formulation the higher the fragility index, the higher the probability of failure.

Following a similar approach, the criticality index, CI_a^y, and the risk index, RI_a^y, are formulated from Eqs. (5.2) to (5.4).

$$\forall_a \, CI_a^y \in CI \subset \mathbb{R} \wedge L \leq CI \leq U \tag{5.2}$$

$$\forall_a RI_a^y \in RI \subset \mathbb{R} \wedge L^2 \leq RI \leq U^2 \tag{5.3}$$

$$RI_a^y = FI_a^y \times CI_a^y \tag{5.4}$$

Both the fragility and criticality indices will be deepened in the next sections.

5.3 Criticality and Fragility Modelling

5.3.1 Deterministic Models

The development of the asset risk index could be based on several methods. This methodological diversity is a result of the asset class variety, data discrepancy and lack of a standardized approach.

The deterministic models are the most applied in the energy industry and are based on the observation of a set of variables, through the development of experiments/tests. The result is determined by the procedure's condition (Meyer 1970). They are, hence, rather engineering-driven, which usually pleases asset experts.

Deterministic models could be used for both asset criticality and fragility assessment. For instance, the criticality index of power transmission lines can be built for current year, $y = i$, as proposed from Eqs. (5.5) to (5.25).

$$CI_a^i = \alpha_{SO}.SO_a^i + \alpha_{SA}.SA_a^i + \alpha_{EN}.EN_a^i \tag{5.5}$$

$$\forall_a CI_a^i \in CI \subset \mathbb{R} \wedge 1 \leq CI \leq 5 \tag{5.6}$$

$$\forall_a SO_a^i \in SO \subset \mathbb{R} \wedge 1 \leq SO \leq 5 \tag{5.7}$$

$$\forall_a SA_a^i \in SA \subset \mathbb{R} \wedge 1 \leq SA \leq 5 \tag{5.8}$$

$$\forall_a EN_a^i \in EN \subset \mathbb{R} \wedge 1 \leq EN \leq 5 \tag{5.9}$$

$$\alpha_{SO}, \alpha_{SA}, \alpha_{EN} \in A \subset \mathbb{R} \wedge 0 \leq A \leq 1 \tag{5.10}$$

where,

SO_a^i refers to the system operation criticality (impact of an asset failure on the power system operation, namely security of supply and reliability);
SA_a^i relates to the impact on people and goods safety;
EN_a^i refers to the impact of an asset failure on the environment;
α_{SO}, α_{SA} and α_{EN} stand for the weights of the operation system, safety and environment criticality indices, respectively.

The attributes of the lines' criticality index are then formulated. The criticality in the system operation perspective, SO^i_a, takes into account the installation type, IT^i_a, quality of service, QS^i_a, security of supply, SS^i_a, average transmitted power, TP^i_a, and the system operation perspective, SP^i_a.

$$SO^i_a = \alpha_{IT} \cdot IT^i_a + \alpha_{QS} \cdot QS^i_a + \alpha_{SS} \cdot SS^i_a + \alpha_{TP} \cdot TP^i_a + \alpha_{SP} \cdot SP^i_a \quad (5.11)$$

where α_{IT}, α_{QS}, α_{SS}, α_{TP} and α_{SP} stand for the weights of the installation type, quality of service, security of supply, average transmitted power and the system operation perspective, respectively.

$$\alpha_{IT}, \alpha_{QS}, \alpha_{SS}, \alpha_{TP}, \alpha_{SP} \in A \subset \mathbb{R} \wedge 0 \leq A \leq 1 \quad (5.12)$$

The attributes of the system operation criticality as formulated as proposed from Eqs. (5.13) to (5.17).

$$\forall_a IT^i_a \in IT \subset \mathbb{N} \wedge 1 \leq IT \leq 5 \quad (5.13)$$

$$\forall_a QS^i_a \in QS \subset \mathbb{N} \wedge 1 \leq QS \leq 5 \quad (5.14)$$

$$\forall_a SS^i_a \in SS \subset \mathbb{N} \wedge 1 \leq SS \leq 5 \quad (5.15)$$

$$\forall_a TP^i_a \in TP \subset \mathbb{N} \wedge 1 \leq TP \leq 5 \quad (5.16)$$

$$\forall_a SP^i_a \in SP \subset \mathbb{N} \wedge 1 \leq SP \leq 5 \quad (5.17)$$

Although the proposed quantitative formulation, the previous attributes are provided through a qualitative scoring procedure [adapted from National Grid et al. (2010)] (Tables 5.1, 5.2 and 5.3).

Regarding the average transmission power, TP^i_a, this model proposes a min-max normalization to the index scale (1 out of 5). Finally, the system operator perspective of each asset criticality, SP^i_a, is given as presented in Table 5.4.

Table 5.1 Scoring for installation type criticality

Installation type	Score
Interconnections, connections to VHV consumers or to single fed DSO	5
Connections to VHV consumers double-fed from cables (main cities)	4
Remaining VHV grid	3
Remaining cables in main cities, connections to railway system, power plants' connections	2
Other circuits	1

Table 5.2 Scoring for quality of service criticality

Quality of service	Score
Substation or VHV consumer loss	5
N − 1 criterion violation, power generation loss, third party supply dependence	4
Violation of security patterns in a N − 2 regime, loss of generation in N − 1	3
Overload of lines connected to power plants	2
Other circuits	1

Table 5.3 Scoring for security of supply criticality

Security of supply	Score
VHV client with load > 100 MW	5
VHV client with load $\leq$ 100 MW	4
Railway system	3
Remaining circuits managed by TSO	2
Circuits operated by DSO	1

Table 5.4 System operator's perspective on the asset criticality

System operator's perspective	Score
Very high criticality	5
High criticality	4
Medium criticality	3
Low criticality	2
Very low criticality	1

When it comes to criticality in the safety of people and goods perspective, SA_a^i, the asset manager should consider both the power line location, LO_a^i, and the failure impact, FL_a^i.

$$SA_a^i = \alpha_{LO} \times LO_a^i + \alpha_{FL} \times FL_a^i \tag{5.18}$$

where α_{LO} and α_{FL} correspond to the weights of the asset location and failure impact, respectively.

$$\alpha_{LO}, \alpha_{FL} \in A \subset \mathbb{R} \wedge 0 \leq A \leq 1 \tag{5.19}$$

The attributes of people and goods safety result from the formulation proposed in Eqs. (5.20) and (5.21).

$$\forall_a LO_a^i \in LO \subset \mathbb{N} \wedge 1 \leq LO \leq 5 \tag{5.20}$$

$$\forall_a FL_a^i \in FL \subset \mathbb{N} \wedge 1 \leq FL \leq 5 \tag{5.21}$$

The location attribute, LO_a^i, is computed through a min-max normalization of km installed in urban zone, whereas the failure impact, FL_a^i, is provided through a qualitative scoring procedure [adapted from National Grid et al. (2010)] (Table 5.5).

To conclude the lines criticality index, one should address the environment criticality, EN_a^i. Similarly to the safety index, the environment criticality gathers the location, EL_a^i, and the failure impact, EF_a^i.

$$EN_a^i = \alpha_{EL} \times EL_a^i + \alpha_{EF} \times EF_a^i \tag{5.22}$$

where α_{EL} and α_{EF} correspond to the weights of the asset location and failure impact, respectively.

$$\alpha_{EL}, \alpha_{EF} \in A \subset \mathbb{R} \wedge 0 \leq A \leq 1 \tag{5.23}$$

The attributes of the environment criticality result from the following formulation.

$$\forall_a EL_a^i \in EL \subset \mathbb{N} \wedge 1 \leq EL \leq 5 \tag{5.24}$$

$$\forall_a EF_a^i \in EF \subset \mathbb{N} \wedge 1 \leq EF \leq 5 \tag{5.25}$$

The location attribute, EL_a^i, is computed through a min-max normalization of km installed in environment protected/sensitive regions, whereas the failure impact, EF_a^i, is provided through a qualitative scoring procedure [adapted from National Grid et al. (2010)] (Table 5.6).

Table 5.5 Scoring for asset failure impact in people and goods safety

Failure impact	Score
Asset failure may result in fatality	5
Asset failure may result in permanently incapacitating injury or considerable goods damage	4
Asset failure may result in reportable injury or goods damage	3
Asset failure may result in minor injury or goods damage	2
Asset failure may have no consequence	1

Table 5.6 Scoring for asset failure impact in the environment

Failure impact	Score
Asset failure may result in an environmental disaster	5
Asset failure may result in a permanent environment impact	4
Asset failure may result in a visible temporary impact	3
Asset failure may result in a minor temporary impact	2
Asset failure may have no consequence	1

Although there is no one-size-fits-all formulation, this model can be adapted to other power and gas assets.

As referred in the beginning of this section, the fragility index can also be built through a deterministic model. The power transformer is probably the most studied electric asset, bearing in mind its value and importance to the utility.

The fragility index, FI_a^i, of a power transformer is formulated considering the asset age, Ag_a^i, oil quality, OQ_a^i, electrical insulation, EI_a^i, performance, Pe_a^i, and operation, Op_a^i.

$$FI_a^i = \alpha_{Ag} \cdot Ag_a^i + \alpha_{OQ} \cdot OQ_a^i + \alpha_{EI} \cdot EI_a^i + \alpha_{Pe} \cdot Pe_a^i + \alpha_{Op} \cdot Op_a^i \tag{5.26}$$

$$\forall_a FI_a^i \in FI \subset \mathbb{R} \wedge 1 \leq FI \leq 5 \tag{5.27}$$

where α_{Ag}, α_{OQ}, α_{EI}, α_{Pe} and α_{Op} stand for the weights of the asset age, oil quality, electrical insulation, performance and operation, respectively.

$$\alpha_{Ag}, \alpha_{OQ}, \alpha_{EI}, \alpha_{Pe}, \alpha_{Op} \in A \subset \mathbb{R} \wedge 0 \leq A \leq 1 \tag{5.28}$$

The attributes that set up the power transformer fragility index are modelled from Eqs. (5.29) to (5.54).

$$\forall_a Ag_a^i \in Ag \subset \mathbb{R} \wedge 1 \leq Ag \leq 5 \tag{5.29}$$

$$\forall_a OQ_a^i \in OQ \subset \mathbb{R} \wedge 1 \leq OQ \leq 5 \tag{5.30}$$

$$\forall_a EI_a^i \in EI \subset \mathbb{N} \wedge 1 \leq EI \leq 5 \tag{5.31}$$

$$\forall_a Pe_a^i \in Pe \subset \mathbb{N} \wedge 1 \leq Pe \leq 5 \tag{5.32}$$

$$\forall_a Op_a^i \in Op \subset \mathbb{R} \wedge 1 \leq Op \leq 5 \tag{5.33}$$

Firstly, the age attribute is modelled via a logistic function.

$$Ag_a^i = \frac{5}{1 + e^{-k(x - x_0)}} \tag{5.34}$$

where,

k refers to the sigmoid steepness;
x stands for the asset age;
x_0 corresponds to the midpoint of the sigmoid (Fig. 5.2).

When it comes to the oil quality attribute, it is computed considering the dissolved gases, DG_a^i, and the degree of polymerization, DP_a^i.

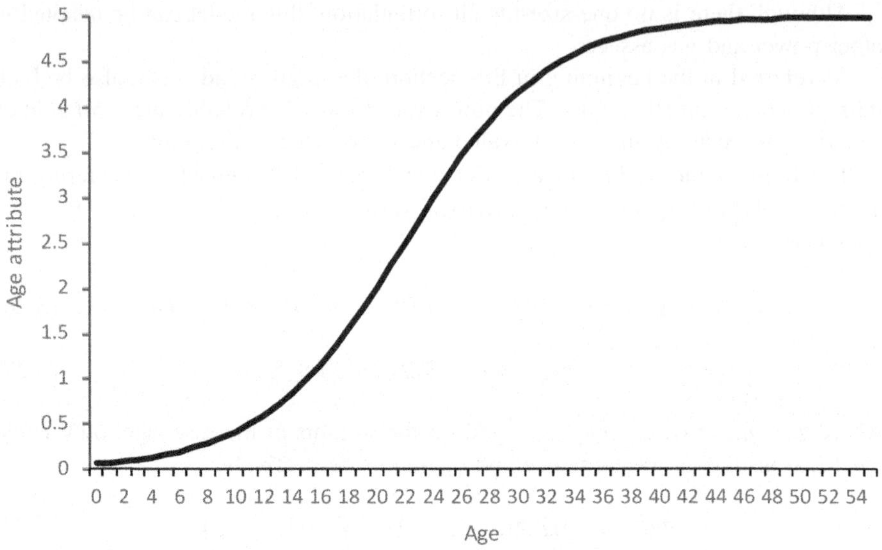

Fig. 5.2 Age attribute contributing to the power transformers fragility

$$OQ_a^i = \alpha_{DG}.DG_a^i + \alpha_{DP}.DP_a^i \tag{5.35}$$

$$\forall_a DG_a^i \in DG \ \subset \mathbb{R} \wedge 1 \leq DG \leq 5 \tag{5.36}$$

$$\forall_a DP_a^i \in DP \ \subset \mathbb{N} \wedge 1 \leq DP \leq 5 \tag{5.37}$$

where α_{DG} and α_{DP} stand for the weights of the dissolved gas attribute and the degree of polymerization attribute, respectively.

$$\alpha_{DG}, \alpha_{DP} \in A \ \subset \mathbb{R} \wedge 0 \leq A \leq 1 \tag{5.38}$$

The dissolved gas analysis is a fundamental inspection to assess the power transformer condition. Indeed, the construction of such an attribute—within the scope of the fragility index—could make use of the industry standards and recommendations for threshold values. One hereby presents an illustrative scoring.

$$DG_a^i = \alpha_{Hi}.Hi_a^i + \alpha_{Me}.Me_a^i + \alpha_{Ac}.Ac_a^i + \alpha_{Et}.Et_a^i + \alpha_{En}.En_a^i + \alpha_{Co}.Co_a^i + \alpha_{Di}.Di_a^i \tag{5.39}$$

In which,

$$Hi_a^i = \begin{cases} 1, & \beta_{H_2} \leq 100 \\ 2, & 100 < \beta_{H_2} \leq 400 \\ 3, & 400 < \beta_{H_2} \leq 700 \\ 4, & 700 < \beta_{H_2} \leq 1800 \\ 5, & \beta_{H_2} > 1800 \end{cases} \tag{5.40}$$

$$Me_a^i = \begin{cases} 1, & \beta_{CH_4} \leq 125 \\ 2, & 125 < \beta_{CH_4} \leq 275 \\ 3, & 275 < \beta_{CH_4} \leq 400 \\ 4, & 400 < \beta_{CH_4} \leq 1000 \\ 5, & \beta_{CH_4} > 1000 \end{cases} \tag{5.41}$$

$$Ac_a^i = \begin{cases} 1, & \beta_{C_2H_2} \leq 2 \\ 2, & 2 < \beta_{C_2H_2} \leq 9 \\ 3, & 9 < \beta_{C_2H_2} \leq 35 \\ 4, & 35 < \beta_{C_2H_2} \leq 80 \\ 5, & \beta_{C_2H_2} > 80 \end{cases} \tag{5.42}$$

$$Et_a^i = \begin{cases} 1, & \beta_{C_2H_4} \leq 50 \\ 2, & 50 < \beta_{C_2H_4} \leq 100 \\ 3, & 100 < \beta_{C_2H_4} \leq 150 \\ 4, & 150 < \beta_{C_2H_4} \leq 200 \\ 5, & \beta_{C_2H_4} > 200 \end{cases} \tag{5.43}$$

$$En_a^i = \begin{cases} 1, & \beta_{C_2H_6} \leq 65 \\ 2, & 65 < \beta_{C_2H_6} \leq 100 \\ 3, & 100 < \beta_{C_2H_6} \leq 150 \\ 4, & 150 < \beta_{C_2H_6} \leq 400 \\ 5, & \beta_{C_2H_6} > 400 \end{cases} \tag{5.44}$$

$$Mo_a^i = \begin{cases} 1, & \beta_{CO} \leq 350 \\ 2, & 350 < \beta_{CO} \leq 700 \\ 3, & 700 < \beta_{CO} \leq 900 \\ 4, & 900 < \beta_{CO} \leq 1400 \\ 5, & \beta_{CO} > 1400 \end{cases} \tag{5.45}$$

$$Di_a^i = \begin{cases} 1, & \beta_{CO2} \leq 2500 \\ 2, & 2500 < \beta_{CO2} \leq 4000 \\ 3, & 4000 < \beta_{CO2} \leq 10,000 \\ 4, & 10,000 < \beta_{CO2} \leq 30,000 \\ 5, & \beta_{CO2} > 30,000 \end{cases} \tag{5.46}$$

$$\alpha_{Hi}, \alpha_{Me}, \alpha_{Ac}, \alpha_{Et}, \alpha_{En}, \alpha_{Mo}, \alpha_{Di} \in A \subset \mathbb{R} \wedge 0 \leq A \leq 1 \tag{5.47}$$

where,

Hi_a^i is the attribute related to hydrogen concentration in the oil (in ppm), β_{H_2}, with weight α_{Hi};
Me_a^i is the attribute related to methane concentration in the oil (in ppm), β_{CH_4}, with weight α_{Me};
Ac_a^i is the attribute related to acetylene concentration in the oil (in ppm), $\beta_{C_2H_2}$, with weight α_{Ac};
Et_a^i is the attribute related to ethylene concentration in the oil (in ppm), $\beta_{C_2H_4}$, with weight α_{Et};
En_a^i is the attribute related to ethane concentration in the oil (in ppm), $\beta_{C_2H_6}$, with weight α_{En};
Mo_a^i is the attribute related to carbon monoxide concentration in the oil (in ppm), β_{CO}, with weight α_{Mo};
Di_a^i is the attribute related to carbon dioxide concentration in the oil (in ppm), β_{CO_2}, with weight α_{Di}.

Given that,

$$\beta_{H_2}, \beta_{CH_4}, \beta_{C_2H_2}, \beta_{C_2H_4}, \beta_{C_2H_6}, \beta_{CO}, \beta_{CO_2} \in \mathbb{R} \tag{5.48}$$

On its side, the degree of polymerization could be formulated as proposed in Eq. (5.49) (Brandtzæg 2015).

$$DP_a^i = \begin{cases} 1, & c_{2FAL} < 0.1 \\ 2, & 0.1 \leq c_{2FAL} < 0.25 \\ 3, & 0.25 \leq c_{2FAL} < 0.5 \\ 4, & 0.5 \leq c_{2FAL} < 1.0 \\ 5, & c_{2FAL} \geq 1.0 \end{cases} \tag{5.49}$$

where $c_{2FAL} \in \mathbb{R}$, standing for the concentration (in ppm) of furaldehyde (2-FAL) in the oil. This furanic compound is a popular variable to assess the degree of polymerization (Martins 2007).

Having developed the attribute referred to the oil quality, one returns to Eq. (5.26) to approach the electrical insulation attribute, EI_a^i, which is based in the power factor measurement (Brandtzæg 2015; Jahromi et al. 2009). The power factor reflects the insulation condition in the power transformer bushings and windings.

$$EI_a^i = \begin{cases} 1, & pf_{\max} < 0.5 \\ 2, & 0.5 \leq pf_{\max} < 0.7 \\ 3, & 0.7 \leq pf_{\max} < 1.0 \\ 4, & 1.0 \leq pf_{\max} < 2.0 \\ 5, & pf_{\max} \geq 2.0 \end{cases} \tag{5.50}$$

where $pf_{max} \in \mathbb{R}$ and refers to the highest measured value of power factor.

As far as the asset performance is concerned, Pe_a^i, the fragility index considers the normalized failure rate of asset a, $FR_a \in \mathbb{R}$, taking into account the min-max normalized failure rate of asset class c, $FR_c \in \mathbb{R}$ (Fig. 5.3).

$$Pe_a^i = \begin{cases} 1, & FR_a < Percentile\,30\ of\ FR_c \\ 2, & Percentile\,30 \leq FR_a < Percentile\,70\ of\ FR_c \\ 3, & Percentile\,70 \leq FR_a < Percentile\,90\ of\ FR_c \\ 4, & Percentile\,90 \leq FR_a < Percentile\,97\ of\ FR_c \\ 5, & FR_a \geq Percentile\,97\ of\ FR_c \end{cases} \tag{5.51}$$

Finally, concerning the operation attribute, Op_a^i, the next formulation includes the operating time for percentile 70 of active power, ON_a^i, as well as the time at overload conditions, OL_a^i, using min-max normalization for the scale of the index (1 out of 5).

$$Op_a^i = \alpha_{ON}.ON_a^i + \alpha_{OL}.OL_a^i \tag{5.52}$$

$$\forall_a Op_a^i \in Op \subset \mathbb{R} \wedge 1 \leq Op \leq 5 \tag{5.53}$$

$$\alpha_{ON}, \alpha_{OL} \in A \subset \mathbb{R} \wedge 0 \leq A \leq 1 \tag{5.54}$$

where α_{ON} and α_{OL} are the weights of attributes ON_a^i and OL_a^i, respectively.

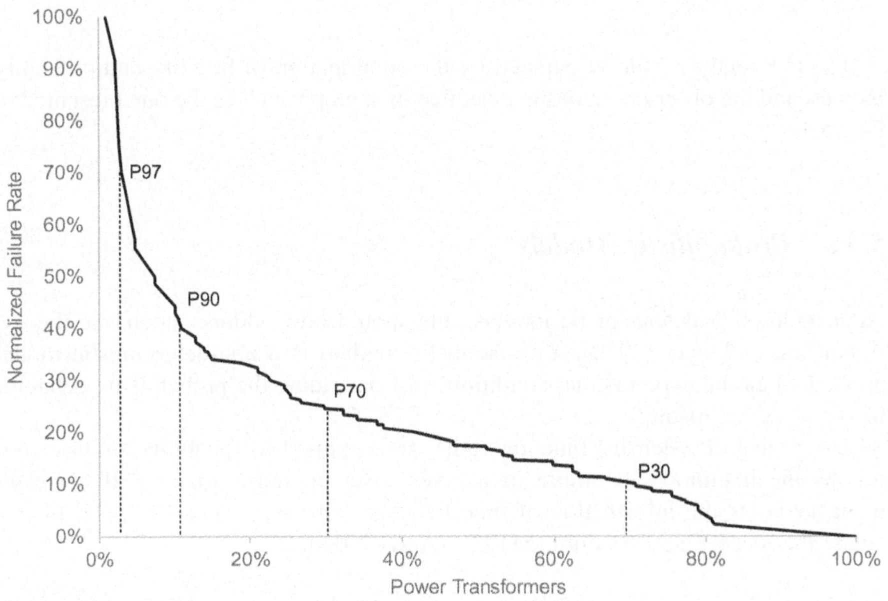

Fig. 5.3 Normalized failure rate of a hypothetical power transformers fleet

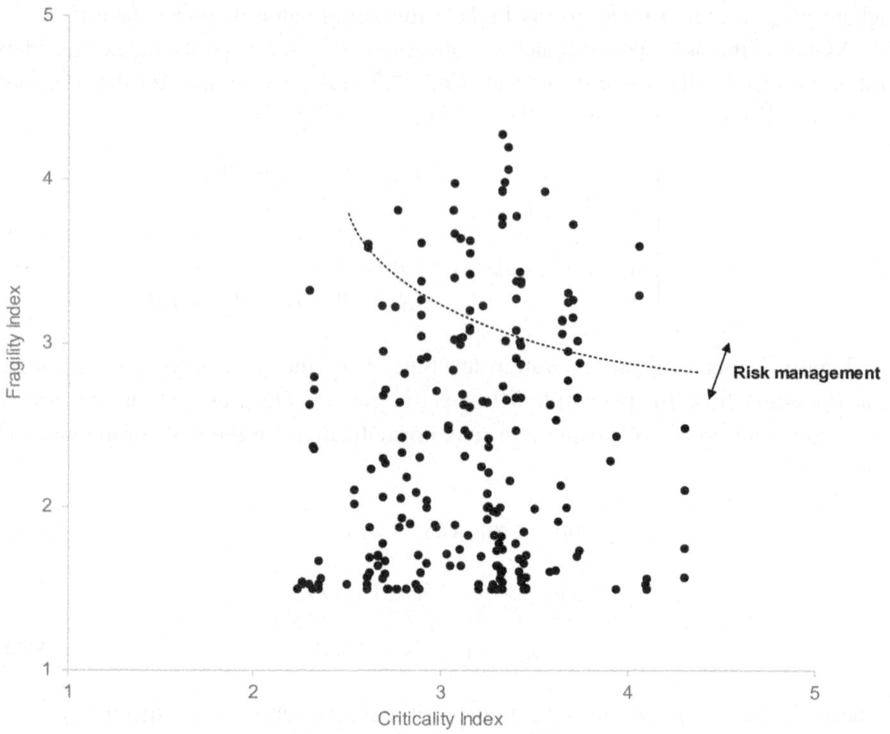

Fig. 5.4 Risk analysis for a theoretical asset fleet

The risk analysis can be pursued by the combination of fragility and criticality indices, and the observation of the asset fleet in a graph such as the one presented in Fig. 5.4.

5.3.2 *Probabilistic Models*

After addressing deterministic models, one should now address stochastic theory. According to Meyer (1970), a probabilistic method is a non-deterministic model applied when the experiments' conditions just determine the probabilistic behavior of the observed result.

The probability density function (pdf), $f(t)$, applied to reliability studies, represents the likelihood of failure of a given asset in time horizon t. If the asset manager holds the information of time between failures, $f(t)$ can be used to calculate the mean time to failure (MTTF) (Assis 2014).

$$MTTF = \int_{0}^{\infty} f(t).t\,dt \tag{5.55}$$

MTTF is adequately referred to non-repairable components, whereas the mean time between failures (MTBF) is related with repairable equipment.

In reliability of engineering systems, it is often investigated the so called "bathtub" curve, which stands for the hazard function of assets, $h(t)$. This curve comprehends three stages: infant (or burn-in) period; maturity (useful life) period; and degradation (wear-out) period. The infant period is characterized by design, manufacturing and commissioning problems, while during the maturity period the hazard function becomes steady (consisting in the failure rate, λ). Finally, in the degradation period the mortality spikes, though the end-of-life can be managed with preventive maintenance and refurbishment (Fig. 5.5).

There is a wide set of pdf worth testing in reliability studies, namely: Weibull; negative exponential; normal; and lognormal. The Weibull function is usually applied to assets subject to degradation phenomena (e.g. corrosion).

$$f(t) = \frac{\alpha}{\beta} \times \left[\frac{t}{\beta}\right]^{\alpha-1} \times e^{-\left(\frac{t}{\beta}\right)^{\alpha}} \tag{5.56}$$

where, α refers to the shape parameter and β stands for the scale parameter (being equal to 1/λ).

The negative exponential function is a particular case of Weibull distribution, when $\alpha = 1$ (and $\beta = MTTF$). This pdf is often applied to casuistic and unexpected failures, but also for complex systems of assets (Assis 2014).

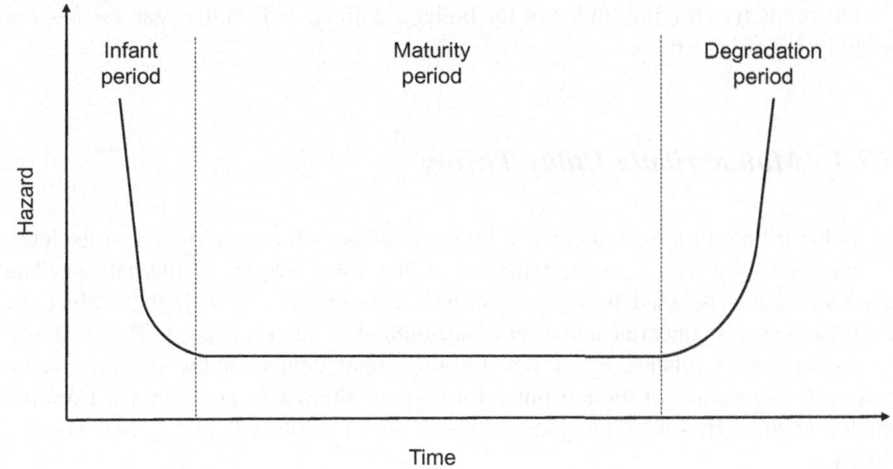

Fig. 5.5 "Bathtub" curve

$$f(t) = \lambda \cdot e^{-\lambda \cdot t} \tag{5.57}$$

The normal function is worth applying when the asset degradation intensifies (Assis 2014). This function is characterized by the mean, μ, and variance, υ.

$$f(t) = \frac{1}{\sigma \cdot \sqrt{2 \cdot \pi}} \cdot e^{-\frac{1}{2} \left[\frac{(t-\mu)}{\sigma}\right]^2} \tag{5.58}$$

When it comes to the lognormal function, it is typically used is cases of very distinct events duration and frequent short-duration events (Assis 2014).

$$f(t) = \frac{1}{\sigma_y \cdot t \cdot \sqrt{2 \cdot \pi}} \cdot e^{-\frac{1}{2} \left[\frac{(\ln(t)-\mu_y)}{\sigma_y}\right]^2} \tag{5.59}$$

where σ_y and μ_y are the mean and standard deviation, respectively, of the natural logarithm of time.

To illustrate the application of probabilistic models to asset management, an example is drawn for the case of a gas boiler. This model attempts to build a predictive fragility index, for year $y = i + t$. The fragility index, FI_a^{i+1}, is developed as a proxy of the probability of failure, given by the negative exponential distribution.

$$\forall_a FI_a^{i+t} \in FI \subset \mathbb{R} \wedge 1 \leq FI \leq 5 \tag{5.60}$$

$$FI_a^{i+t} = \left(\lambda_a . e^{-(\lambda_a . (i+t))}\right) \times 5 \tag{5.61}$$

where $\lambda_a \in \mathbb{R}$ is the average failure rate of boiler a.

The predictive fragility index of the boiler a with $\lambda_a = 1$, in the year $y = i + 1$ is roughly 1.8 (Fig. 5.6).

5.3.3 *Multiattribute Utility Theory*

In multiattribute problems there is a finite set of predefined alternatives (discrete), namely $A = \{a_1, a_2, \ldots, a_m\}$, where A is the finite set m of alternatives. The alternatives are assessed by a set of criteria $C = \{c_1, c_2, \ldots, c_n\}$, that reflect the decision maker's concerns, and a set of attributes $X = \{x_{1,1}, \ldots, x_{m,n}\}$. The attributes levels serve as yardstick by which the attainment degree of the criteria can be assessed. The values of the attributes for a given alternative are obtained from the model or through subjective judgements (Catrinu 2006; Chankong and Haimes 2008).

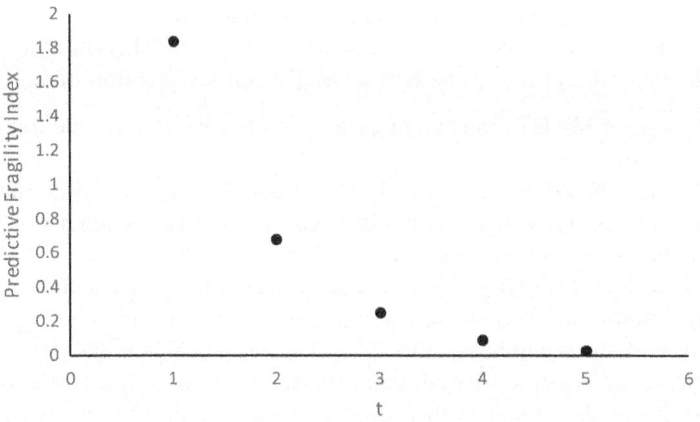

Fig. 5.6 Predictive fragility index for a gas boiler with $\lambda_a = 1$

There is a wide range of methods for solving multiattribute problems. One of the most applied approaches is based on the multi-attribute utility theory (MAUT) and assumes that the decision maker is able to specify precise answers to a wide range of preference elicitation questions. There is a set of methods that follow this approach, such as the direct rating and mid-point methods for constructing value functions, the Analytical Hierarchy Process method, Simple Multi-Attribute Rating Technique, and the Utility Theory Additive. Multi-attribute utility theory, proposed by Keeney and Raiffa (1976), assumes that the decision maker's preference "can be quantified, measured, and represented in the form of a real-valued function called multi-attribute value function or utility function" (Chankong and Haimes 2008). With this utility function, $u_i(x)$, the decision problem is solved through a routine evaluation and search procedure. Thus, having identified the attributes and alternatives, the next tiers should be followed, as recommended by Chankong and Haimes (2008): (1) verify the existence of the multi-attribute value function; (2) select a suitable form of the function; (3) construct component functions to use in step 2; (4) determine scaling constants; and (5) check consistency and make the final analysis.

Therefore, utility theory attempts to construct the preference order by directly eliciting the decision maker's preference. Risk behavior is linked to how the decision maker values a given attribute, in face of uncertainty. If a decision maker judges the actual values of an attribute, in situations of uncertainty, in the same way as under certainty, such a decision maker is described as risk-neutral. Chankong and Haimes (2008) defined that a decision maker is said to be:

i. Risk neutral if and only if $u_i(x)$ is a linear function of x_i;
ii. Risk averse if and only if $u_i(x)$ is a strictly concave function of x_i;
iii. Risk prone if and only if $u_i(x)$ is a strictly convex function of x_i.

In the scope of MAUT, one hereby proposes a formulation for the fragility index of overhead lines.

Firstly, one should have in mind that although common asset management approaches address OHL from a circuit-based perspective, an atomic vision of the asset will lead to a more accurate fragility index (Fig. 5.7).

The circuit-based fragility index is usually preferred by maintenance engineers, since the amount of data of an atomic vision is considerably higher. For a 200-circuit grid, there will be roughly 20,000 spans. Yet a granular view of the asset will enable actual condition-based maintenance, since there is a fragility index per atom throughout the circuit (which can have more than 100 km length). In this model each atom is a unique combination of the supports and spans.

The formulation of atomic fragility index starts with the identification and modelling of the OHL criteria, c, and attributes, x.

c_1: Ageing:

$x_{1,1}$: support age;
$x_{1,2}$: span age.

c_2: Location:

$x_{2,1}$: corrosion;
$x_{2,2}$: altitude.

c_3: Performance:

$x_{3,1}$: failure rate.

c_4: Maintenance:

$x_{4,1}$: OPEX of the last maintenance action;
$x_{4,2}$: years since the last maintenance action.

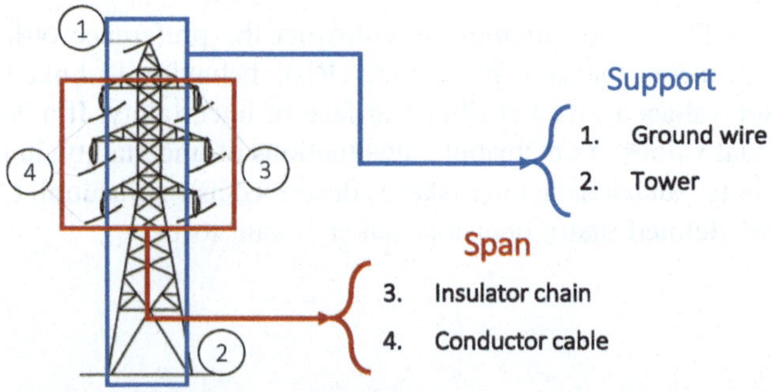

Fig. 5.7 Atomic fragility index for OHL

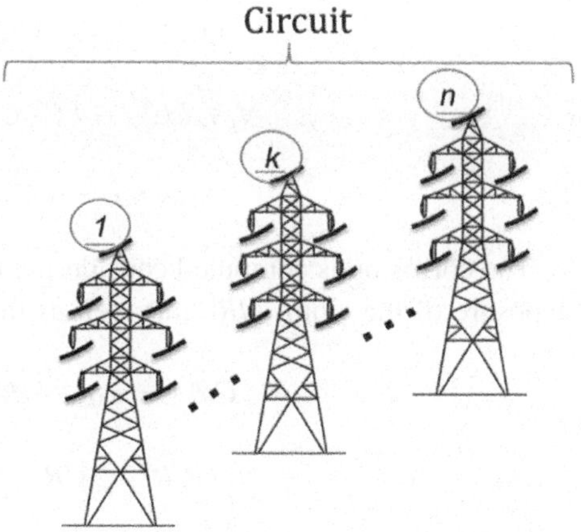

Fig. 5.8 OHL illustrative circuit

c_5: Operation:

$x_{5,1}$: active power transmission.

To keep the experts' preferred view (circuit-based), the highest atomic fragility index of atom a in circuit j, $FI^y_{j,a}$, gives the score to the circuit fragility index, FI^y_j, as follows (Fig. 5.8).

$$FI^y_j = max(FI^y_{j,a}) \tag{5.62}$$

$$\forall_j FI^y_j, \forall_j \forall_a FI^y_{j,a} \in FI \subset \mathbb{R} \wedge 1 \leq FI \leq 3 \tag{5.63}$$

For a fragility index focused in current asset condition, let $y = i$.

Starting to grip the ageing criterion, $AG^i_{j,a}$, this model encompasses both the support age, $SU^i_{j,a}$, and span age, $SP^i_{j,a}$, of atom a at circuit j.

$$AG^i_{j,a} = \begin{cases} 1, & \frac{SU^i_{j,a} + SP^i_{j,a}}{2} \leq 5 \\ 2, & 5 < \frac{SU^i_{j,a} + SP^i_{j,a}}{2} \leq 30 \\ 3, & 30 < \frac{SU^i_{j,a} + SP^i_{j,a}}{2} \end{cases} \tag{5.64}$$

where,

$$\forall_j \forall_a SU^i_{j,a}, \forall_j \forall_a SP^i_{j,a} \in W \subset \mathbb{R}_+ \tag{5.65}$$

As far as location is considered, $LC^i_{j,a}$, the fragility index model proposes the weighing of corrosion, $CR^i_{j,a}$, and altitude, $AL^i_{j,a}$.

$$LC^{i}_{j,a} = \alpha_{CR} \cdot CR^{i}_{j,a} + \alpha_{AL} \cdot AL^{i}_{j,a} \tag{5.66}$$

$$\forall_j \forall_a LC^{i}_{j,a} \in LC \subset \mathbb{R} \wedge 1 \le LC \le 3 \tag{5.67}$$

$$\alpha_{CR}, \alpha_{AL} \in A \subset \mathbb{R} \wedge 0 \le A \le 1 \tag{5.68}$$

The corrosion is computed considering two main variables, namely the maritime exposure of the atom, $MR^{i}_{j,a}$, as well as the pollution exposure, $PO^{i}_{j,a}$.

$$CR^{i}_{j,a} = \alpha_{MR} \cdot MR^{i}_{j,a} + \alpha_{PO} \cdot PO^{i}_{j,a} \tag{5.69}$$

$$\forall_j \forall_a CR^{i}_{j,a} \in CR \subset \mathbb{R} \wedge 1 \le CR \le 3 \tag{5.70}$$

$$\alpha_{MR}, \alpha_{PO} \in A \subset \mathbb{R} \wedge 0 \le A \le 1 \tag{5.71}$$

where α_{MR} and α_{PO} are the weights of attributes $MR^{i}_{j,a}$ and $PO^{i}_{j,a}$, respectively.

The maritime exposure includes the atom distance to shore, d_a, and the exposure to salty environment, s_a. Firstly, the salt exposure is clustered into four geolocated groups, i.e. A, B, C and D (according to data from the national weather agency). Given the OHL atom coordinates, the model plots the atom into one of those clusters. On its side, each of the clusters provides a specific corrosion curve, as shown in Fig. 5.9. Considering the curve assigned to a specific atom and its distance to shore, the maritime exposure is then calculated.

$$MR^{i}_{j,a} = \begin{cases} 3, & cluster\,A \wedge d_a \in [0; \infty) \\ \frac{ln(d_a - 41.5)}{\ln(37)} + 0.59, & cluster\,B \wedge d_a < 40 \\ \frac{ln(d_a - 21.5)}{\ln(17)} + 1.02, & cluster\,C \wedge d_a < 21 \\ \frac{ln(d_a - 10)}{\ln(5.5)} + 1.34, & cluster\,D \wedge d_a < 10 \\ 0, & otherwise \end{cases} \tag{5.72}$$

when it comes to the pollution exposure, $PO^{i}_{j,a}$, this model proposes to include the industrial pollution, IP_a, and relative humidity, RH_a, weighed through α_{IP} and α_{RH}, respectively.

$$PO^{i}_{j,a} = \alpha_{IP} \cdot IP_a + \alpha_{RH} \cdot RH_a \tag{5.73}$$

$$\forall_j \forall_a PO^{i}_{j,a} \in PO \subset \mathbb{R} \wedge 1 \le PO \le 3 \tag{5.74}$$

$$IP_a = \begin{cases} 1, & c_{IP} \le 10 \\ 2, & 10 < c_{IP} \le 75 \\ 3, & 75 < c_{IP} \end{cases} \tag{5.75}$$

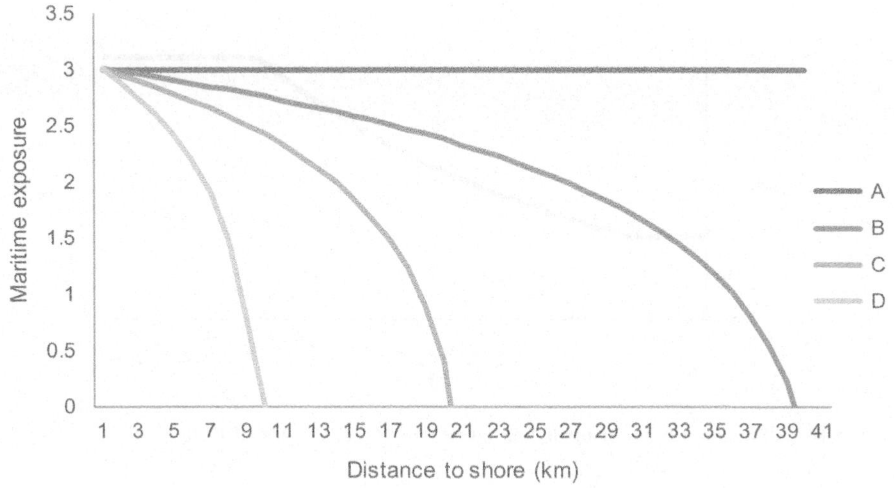

Fig. 5.9 Maritime exposure for modelling the atomic fragility index

$$RH_a = \begin{cases} 1, & c_{RH} \leq 70 \\ 2, & 70 < c_{RH} \leq 85 \\ 3, & 85 < c_{RH} \end{cases} \quad (5.76)$$

$$\alpha_{IP}, \alpha_{RH} \in A \subset \mathbb{R} \wedge 0 \leq A \leq 1 \quad (5.77)$$

where $c_{IP} \in \mathbb{R}$ and $c_{RH} \in \mathbb{R}$ stand for the concentration of industrial pollutants, such as SOx (in mg/m^2), and the relative humidity (in %), respectively.

Regarding the altitude attribute of OHL atom a of circuit j, $AL^i_{j,a}$, this model proposes the formulation provided in Eq. (5.78), where h_a refers to the altitude in meters (Fig. 5.10).

$$AL^i_{j,a} = \begin{cases} 1, & h_a \leq 100 \\ \frac{e^{(h_a-300)*\frac{ln(2)}{300}}}{2} \times 3, & 100 < h_a \leq 600 \\ 3, & h_a > 600 \end{cases} \quad (5.78)$$

As far as the asset performance is concerned, $PE^i_{j,a}$, the fragility index considers the normalized failure rate of OHL atom a, $FR_{j,a} \in \mathbb{R}$, taking into account the min-max normalization of failure rate of the whole OHL atoms, $FR_A \in \mathbb{R}$.

$$PE^i_{j,a} = \begin{cases} 1, & FR_{j,a} < Quartile\, 1 \; of \; FR_A \\ 2, & Quartile\, 1 \leq FR_{j,a} < Quartile\, 3 \; of \; FR_A \\ 3, & FR_{j,a} \geq Quartile\, 3 \; of \; FR_A \end{cases} \quad (5.79)$$

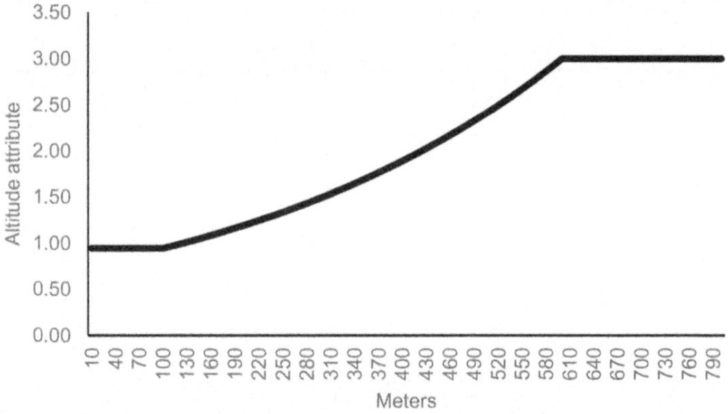

Fig. 5.10 Altitude attribute for modelling the atomic fragility index of OHL

Concerning the operation attribute, $OP^i_{j,a}$, the next formulation includes the operating time for percentile 70 of active power, $ON^i_{j,a}$, as well as the time at overload conditions, $OL^i_{j,a}$, using min-max normalization for the scale of the index (1 out of 3).

$$OP^i_{j,a} = \alpha_{ON}.ON^i_{j,a} + \alpha_{OL}.OL^i_{j,a} \tag{5.80}$$

$$\forall_j \forall_a OP^i_{j,a} \in OP \subset \mathbb{R} \land 1 \leq OP \leq 3 \tag{5.81}$$

$$\alpha_{ON}, \alpha_{OL} \in A \subset \mathbb{R} \land 0 \leq A \leq 1 \tag{5.82}$$

where α_{ON} and α_{OL} are the weights of attributes $ON^i_{j,a}$ and $OL^i_{j,a}$, respectively.

Finally, the maintenance attribute, $MA^i_{j,a}$, is developed by considering the time (in years), t_m, since the last action and its present value, $PV^i_{j,a}$. The OPEX present value is computed as proposed in Eq. (5.83).

$$PV^i_{j,a} = Mc_{j,a} \times \varphi_{t_m} \tag{5.83}$$

$$\varphi_{t_m} = \mathrm{e}^{-1/\lambda \cdot t_m} \tag{5.84}$$

where $Mc_{j,a}$ refers to the OPEX of the last maintenance action in OHL atom a of circuit j, and φ_{t_m} provides the maintenance valorization over time (Fig. 5.11).

After computing the OPEX present value, the model applies a normalization of $PV^y_{j,a}$, referred $Nm_{j,a}$, taking into account the min-max normalization of OPEX present value of the whole OHL atoms, Nm_A. Finally, the maintenance attribute, $MA^i_{j,a}$, is computed as presented in Eq. (5.85).

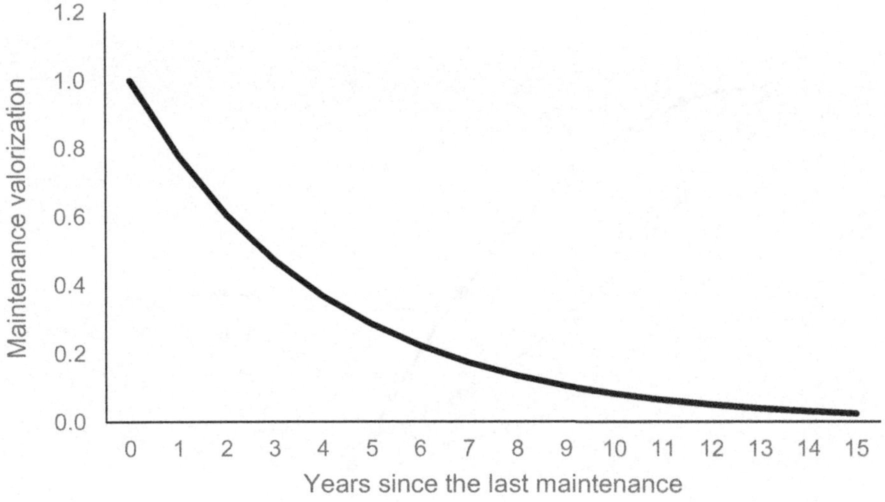

Fig. 5.11 Maintenance valorization over time

$$MA^{i}_{j,a} = \begin{cases} 3, & Nm_{j,a} < Quartile\,1\ of\ Nm_A \\ 2, & Quartile\,1 \le Nm_{j,a} < Quartile\,3\ of\ Nm_A \\ 1, & Nm_{j,a} \ge Quartile\,3\ of\ Nm_A \end{cases} \tag{5.85}$$

Having modelled and calculated the attributes, one proposes applying the utility theory to engage the asset manager in a non-prescriptive way. Utility (or value) functions require: verifying assumptions; construction of the individual utility functions; and indifference judgments to build the multiattribute value function (Matos 2005).

For the case of the OHL fragility index, the utility theory is applied by the following methodology.

1. An expert gives a fragility index score to each circuit, taking into account his/her deep understanding of this asset class, and the analysis of the available dataset.
2. Calculation of the distance, ΔFI^{i}_{j}, between the fragility index of each attribute and the one scored by the expert for circuit j.
3. Construction of value functions for each attribute, considering the ΔFI^{i}_{j} (the lower the distance, the larger the utility).
4. Determination of the parameters k that characterize the value function.
5. Calculation of the fragility index for each circuit, applying the utility theory.

The construction of the value functions is based on the examination of the ΔFI^{i}_{j} for each attribute (e.g. maximum ΔFI^{i}_{j} from all the circuits), bearing in mind that the convexity of the curve reflects the expert confidence. From example depicted in Fig. 5.12, it is possible to infer that the expert relies on the attributes related to age

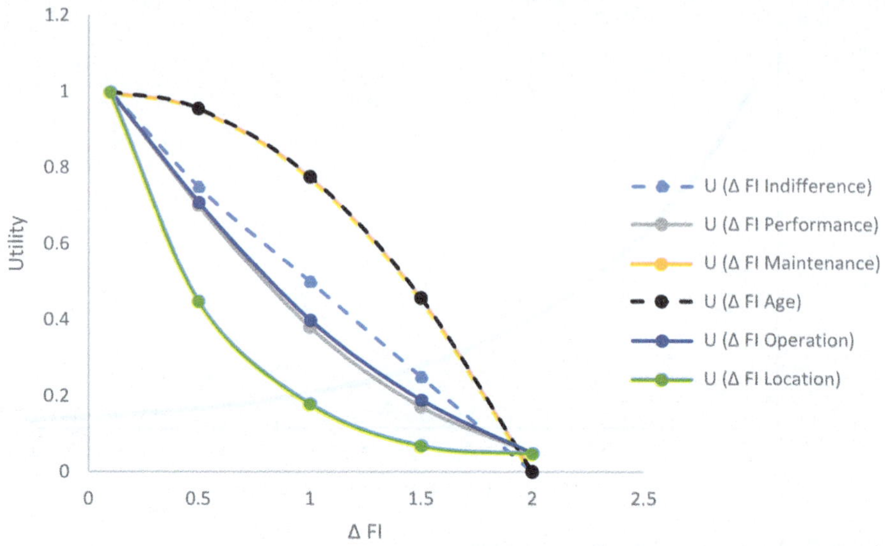

Fig. 5.12 Value functions for each fragility index attributes

and maintenance, showing on the other hand limited confidence on the location contribution for the fragility index.

The construction of the multiattribute value function, v, implies the determination of the k of each attribute. For the present example a general additive value function was considered, while the individual value functions are illustrated in Fig. 5.12.

$$v = k_{AG}.v_{AG} + k_{LC}.v_{LC} + k_{PE}.v_{PE} + k_{MA}.v_{MA} + k_{OP}.v_{OP} \tag{5.86}$$

Considering that the sum of the parameters is the unity, determination of k's is straightforward. It is possible to define two pairs of points, each point on the same indifference curve, which are valued by the expert.

For instance, the expert judges two "extreme" points of operation and location value functions: P (0.5; 0.5); Q (1.5; 1.5). The expert is either indifferent or prefers one of the points. According to the expert, P $\succcurlyeq$ Q.

Then, P is kept equal and Q is improved, i.e.: Q' (1.0; 1.5). The expert still prefers P, that is to say P $\succcurlyeq$ Q'. Again, P is maintained and Q' is improved as follows: Q" (0.5; 1.5).

Now, the expert is indifferent, so P $\sim$ Q and k_{LC} and k_{Op} are obtained as follows.

$$k_{OP} = \frac{1}{1 + \upsilon_{OP}} \tag{5.87}$$

$$k_{LC} = 1 - k_{OP} \tag{5.88}$$

With the calculation of all the parameters of the multiattribute utility function, the asset manager is able to obtain the fragility index of each OHL circuit, by engaging the asset expert while avoiding weight prescription.

5.3.4 Learning Algorithms

Learning algorithms consist in a promising approach to handle predictive maintenance problems. These are grouped in the next categories: supervised models; and unsupervised models.

The supervised models can be either classification (finite and discrete dataset) or regression (infinite and continuous dataset) problems.

There is a set of methods that fall into this category (Gama et al. 2017).

- Distance-based methods: nearest neighbor and its variants;
- Probabilistic methods: naive Bayes and Bayesian networks;
- Search-based algorithms: decision trees and regression trees;
- Optimization methods: artificial neural networks; deep learning and support vector machines;
- Ensemble methods: bagging technique (random forests) and boosting technique (gradient boosting).

The unsupervised models, on their side, provide the identification of intrinsic data properties, in order to build representations that support the decision making process. There are three main techniques used in unsupervised models: summarization; association; and clustering.

The most analyzed clustering methods are presented next (Gama et al. 2017):

- hierarchical algorithms;
- square error partitioning algorithms (being the k-means the most applied);
- density-based algorithms;
- graph-based algorithms;
- neural networks based algorithms;
- grid based algorithms.

In the context of asset management problems, if an electricity utility seeks to build the fragility index of circuit breakers, the next methodology can be developed.

1. Identification of the asset condition variables:

   <u>Asset registry</u>

   - Age;
   - Technology type;
   - Type of application;
   - Location (substation);

Performance and Maintenance

- Forced and Fault Outages;
- Number of maintenance orders (corrective and preventive);
- Voltage and restrike per operation;
- Tripping current per operation (number, value and type of tripping currents per equipment);
- Let-through energy (I^2T) per operation;
- Duration of the short-circuit per operation;

Operation

- Number of equipment maneuvers and date of the last maneuver;
- Operating time delay discrepancy among the circuit breaker's poles per operation;
- Power, voltage and current measures;

Inspection data

- SF_6 condition;
- SF_6 leakage rate;
- Contact resistance;
- Thermography results (hot spots).

2. Data normalization, variable reduction and creation of synthetic data (to prevent overfitting and increase the algorithm accuracy):

   - Age;
   - Type of application;
   - Number of failures;
   - Number of maneuvers;
   - Let-through energy (I^2T) per operation;
   - Percentile 75 of the operating current.

3. Calculation of the fragility index per attribute, in a specified scale (e.g. 1 out of 5).
4. Expert judgement for the global fragility index of each circuit breaker
5. Training and testing different learning algorithms.
6. Implementation of the preferred algorithm (Fig. 5.13).

For example, the data scientist can start by applying a supervised model, specifically a classification approach (bearing in mind that the expert's judgement is discrete). Figure 5.14 sketches the results of the application of the gradient boosting classifier. One can rapidly infer that this algorithm lacks predictability ($R^2 = 0.351$), since there is a considerable error between the predicted fragility index and the expert's judgement.

To improve the previous results, the data scientist can increase the training set, in order to enhance the algorithm training. Figure 5.15 shows progress in the algorithm's coefficient of determination ($R^2 = 0.686$), though the algorithm is unable to

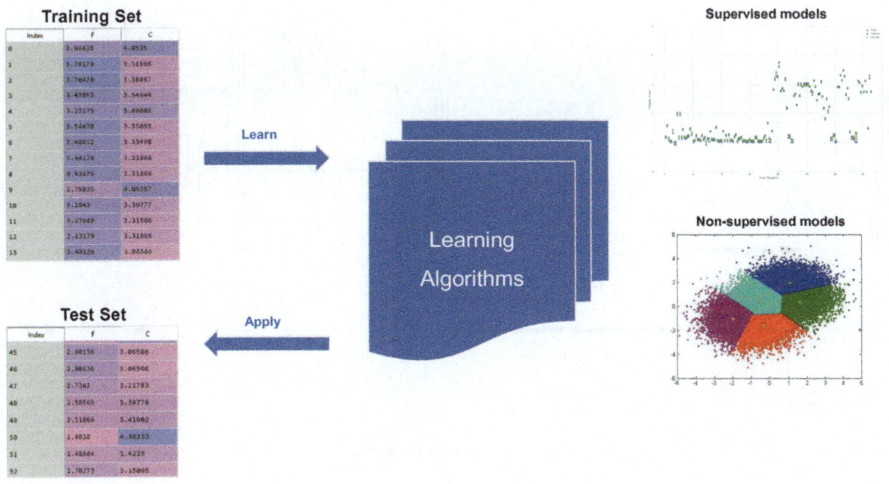

Legend: F – asset fragility index; C – asset criticality

Fig. 5.13 Architecture of machine learning applied to asset management

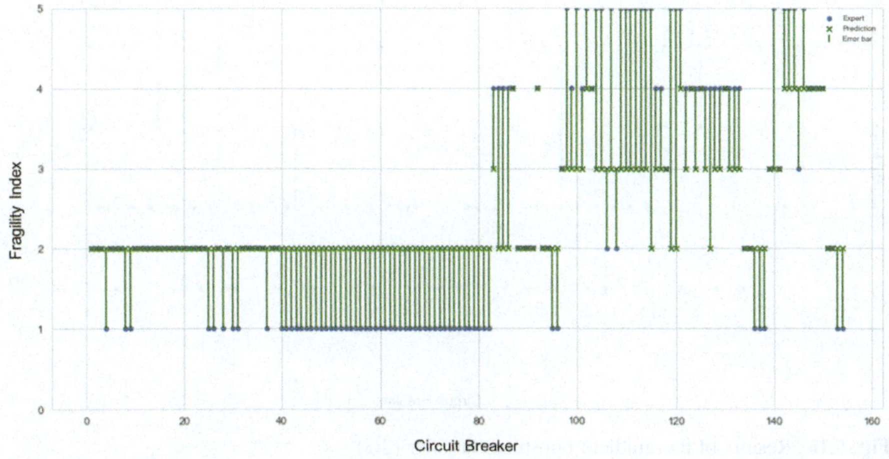

Fig. 5.14 Results of the gradient boosting classifier (1/3)

correctly evaluate circuit breakers with expert's fragility index equal to 5 (highest fragility).

The data scientist can now try to increase (randomly) the diversity of fragility indices, within the training set. From Fig. 5.16 it is possible to observe that the algorithm eventually learned with the expert's perception, together with the available dataset ($R^2 = 0.831$). These results validate the variables reduction process and reinforces the expert's confidence in the model.

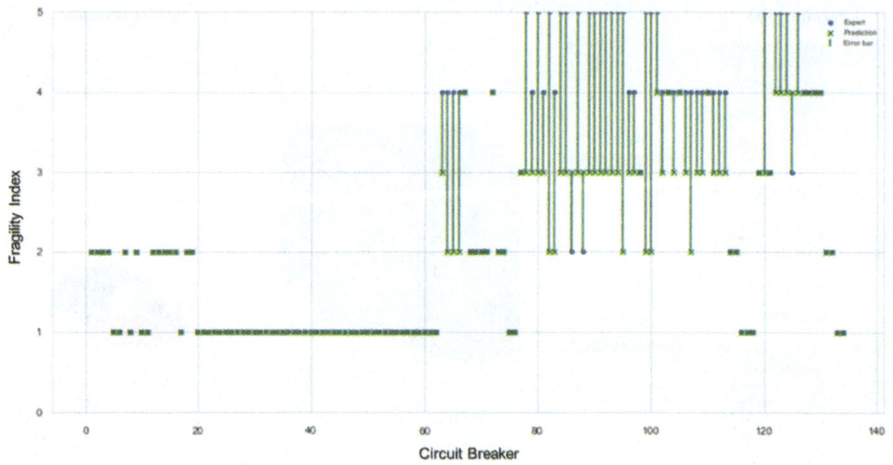

Fig. 5.15 Results of the gradient boosting classifier (2/3)

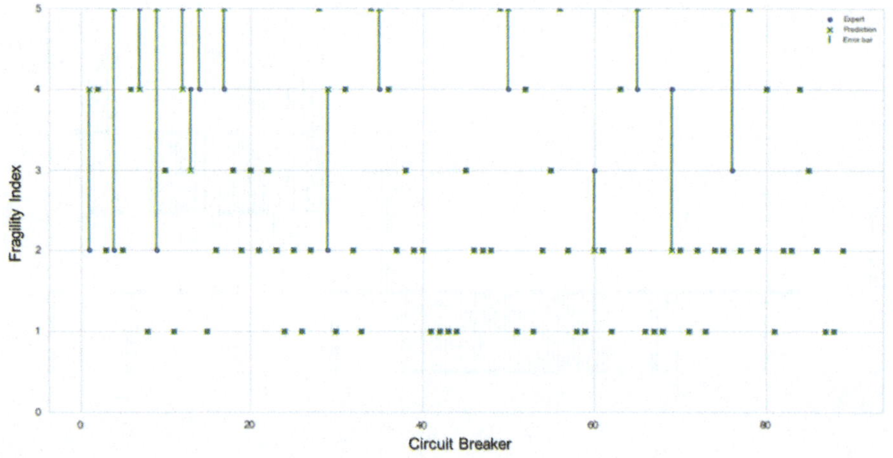

Fig. 5.16 Results of the gradient boosting classifier (3/3)

Additionally, it is also possible to conclude that the algorithm led to a rather conservative fragility index. Figure 5.17 demonstrates that the analytical model results in more circuit breakers with fragility index equal to five, than from the expert's perception. From an asset management perspective, a conservative fragility index is rather adequate than an optimistic algorithm.

This model was developed targeting the asset's current condition, i.e. fragility index. Yet this approach could also be worth developing for multi-year prediction. Such a long-term prediction is undoubtedly useful for replacement planning.

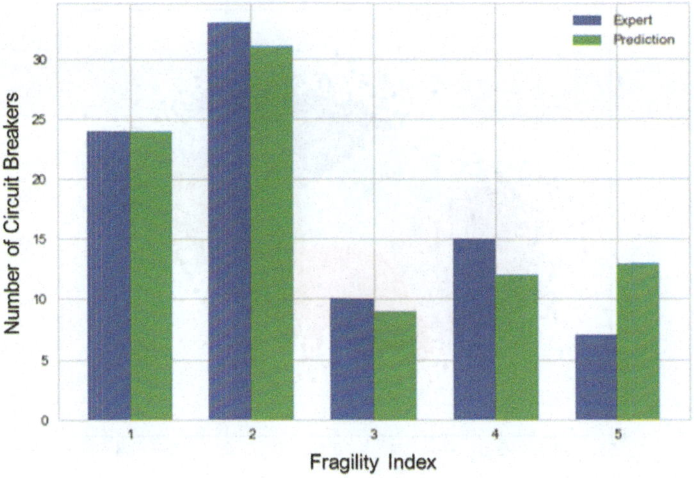

Fig. 5.17 Comparison between the model and expert fragility indices

The FI_a^{i+t} model (where t stands for the incremental years) is implemented by projecting each attribute for the target year, via deterministic or probabilistic approaches. The algorithm is then applied to the new dataset, foreseen for the year $y = i + t$.

Besides the supervised model, the data scientist can evaluate the suitability of an unsupervised model, namely a clustering algorithm. One of the most applied clustering algorithms consists in the k-means. This algorithm seeks to cluster the data points, by calculating iteratively centroids (i.e. mean of each cluster). The k-means algorithm optimizes the number, k, and positioning of centroids, and then populates all the data points into the nearest cluster (while constraining the number of centroids) (Figs. 5.18 and 5.19).

Table 5.7 provides an example of insights that can be inferred from the concerned clustering algorithm. Although this approach does not lead to a numerical fragility index, it provides a useful organization of the assets. For instance, each cluster can be subject to a specific maintenance strategy. In addition, this model is worth applying when there is lack of internal expertise to build a supervised model. A data scientist (who is not an asset specialist) swiftly concludes that cluster 1 is the most “fragile” set of circuit breakers, whereas cluster 2 gathers assets with lower fragility index.

As a drawback, this model does not provide a direct prioritization metric and the company experts might disbelieve on a non-analytical methodology.

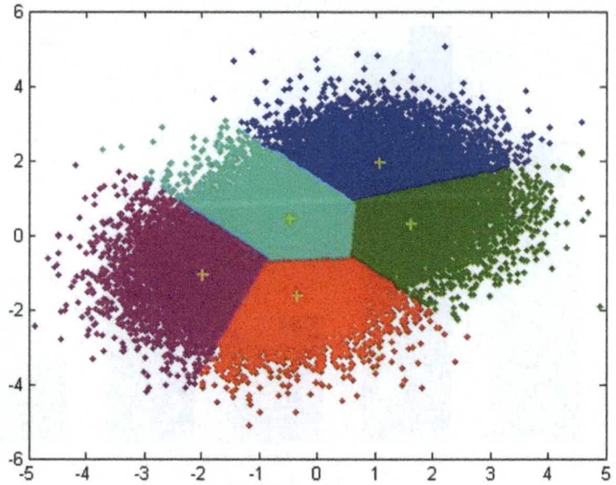

Fig. 5.18 Clustering algorithm representation (Cao 2008)

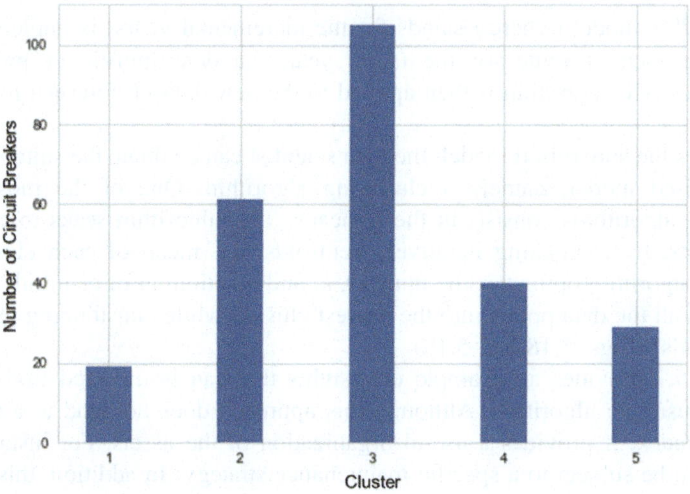

Fig. 5.19 Result of a clustering algorithm

Table 5.7 Clusters of circuit breakers attributes

Cluster	Age	Let-through energy (I^2T)	Current (P75)	Failure rate	Maneuvers
1	High	High	Mid	All	All
2	Low	Low	Low	Low	Low
3	High	Low	Mid	Low	High
4	Med	Low	Mid	High	All
5	High	Mid	All	Low	All

References

Assis R (2014) Apoio à Decisão em Manutenção na Gestão de Ativos Físicos, Lidel

Belton V, Stewart T (2002) Multiple criteria decision analysis—an integrated approach. Kluwer Academic Publishers, Dordrecht

Brandtzæg G (2015) Health indexing of norwegian power transformers, master of energy and environmental engineering. Norwegian University of Science and Technology, Department of Electric Power Engineering, Trondheim

Cao Y (2008) Efficient K-means clustering using JIT. Mathworks, file exchange

Catrinu M (2006) Decision aid for planning local energy systems—application of multi-criteria decision analysis. Norwegian University of Science and Technology

Chankong V, Haimes Y (2008) Multiobjective decision making—theory and methodology. Dover Publications, USA

Clímaco J (1981) Programação Matemática com Objectivos Múltiplos, Dissertação de Doutoramento. University of Coimbra, Coimbra

Figueira, J, Grecco S, Ehrgott M (2005) Multiple criteria decision analysis: state of the art surveys. Springer's International Series, Berlin

Gama J, Carvalho APdL, Faceli K, Lorena AC, Oliveira M (2017) Extração de Conhecimento de Dados—Data Mining, Edições Sílabo, Lda

Hwang C, Masud A (1979) Lecture notes in economics and mathematical systems

Jahromi A, Piercy R, Cress S, Service J (2009) An approach to power transformer asset management using health index. IEEE Electr Insul Mag 25(2):20–34

Keeney R (1992) Value-focused thinking. a path to creative decision making. Harvard University Press, Cambridge

Keeney R, Raiffa H (1976) Decisions with multiple objectives: preferences and value tradeoffs. Wiley, New York

Martins MAG (2007) Furfuraldeído—Um Indicador Prático Da Degradação Térmica Do Papel Kraft De Transformadores. *Ciência e Tecnologia dos Materiais* 19(1/2)

Matos M (1988) Ajuda à Decisão Multicritério—Novas Contribuições, Dissertação de Doutoramento. Faculty of Engineering, University of Porto

Matos MA (2005) Decisions under uncertainty, decision aid methodologies, class notes. Faculty of Engineering, University of Porto, Porto

Meyer PL (1970) Introductory probability and statistical applications. Addison-Wesley Publishing Company, Boston

National Grid, SP Transmission Limited, Scottish Hydro Electric Transmission Limited (2010) Network output measures methodology

Roy B (1993) Decision science or decision-aid science? Eur J Oper Res 66:184

Roy B (1996) Multicriteria methodology for decision aiding, Nonconvex optimization and its applications. Kluwer Academic Publishers, Dordrecht

Taleb NN (2012) Antifragile: things that gain from disorder. Random House, New York

The manufacturer's authorised representative in the EU is Springer
Nature Customer Service Centre GmbH, Europaplatz 3, 69115 Heidelberg,
Germany. If you have any concerns regarding our products, please
contact ProductSafety@springernature.com

Printed and bound by CPI Group (UK) Ltd, Croydon, CR0 4YY
30/09/2026
02225472-0018